Series on Complexity Science – Vol. 4

A Guide to Temporal Networks

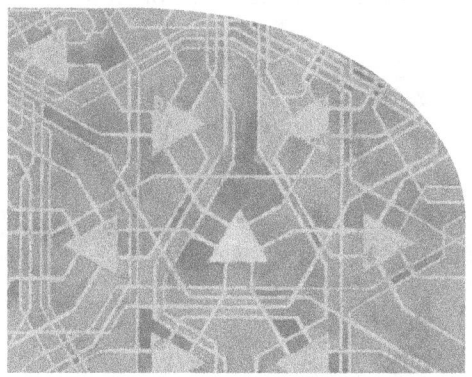

Series on Complexity Science

ISSN: 1755-7453

Series Editor: Henrik Jeldtoft Jensen *(Imperial College London, UK)*

Series on Complexity Science – Vol. 4

A Guide to Temporal Networks

Naoki Masuda
University of Bristol, UK

Renaud Lambiotte
University of Namur, Belgium

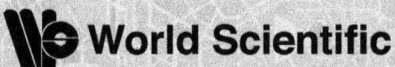
World Scientific

NEW JERSEY · LONDON · SINGAPORE · BEIJING · SHANGHAI · HONG KONG · TAIPEI · CHENNAI · TOKYO

Published by

World Scientific Publishing Europe Ltd.

57 Shelton Street, Covent Garden, London WC2H 9HE

Head office: 5 Toh Tuck Link, Singapore 596224

USA office: 27 Warren Street, Suite 401-402, Hackensack, NJ 07601

Library of Congress Cataloging-in-Publication Data

Names: Masuda, Naoki, 1976–   author. | Lambiotte, Renaud, author.

Title: A guide to temporal networks / Naoki Masuda (University of Bristol, UK),
  Renaud Lambiotte (University of Namur, Belgium).

Other titles: World Scientific series on complexity science ; v. 4.

Description: Covent Garden, London : Imperial College Press, [2016] |

Series: Series on complexity science ; vol. 4 | Includes bibliographical references and index.

Identifiers: LCCN 2016015327| ISBN 9781786341143 (hardcover ; alk. paper) |
  ISBN 178634114X (hardcover ; alk. paper)

Subjects: LCSH: System analysis. | Computational complexity. |
  Social networks--Mathematical models.

Classification: LCC QA402 .M3585 2016 | DDC 003--dc23

LC record available at https://lccn.loc.gov/2016015327

British Library Cataloguing-in-Publication Data

A catalogue record for this book is available from the British Library.

Preface

What do the human brain, social systems, the World Wide Web and metabolism have in common? They are classical examples of complex systems, i.e., systems composed of a large number of relatively simple elements in interaction and exhibiting emerging collective phenomena. As Herbert Simon wrote in "The Architecture of Complexity" half a century ago, "it may not be entirely vain to search for common properties among diverse kinds of complex systems". This search requires a modelling language to describe systems of different types in a unified framework. One such language is that of networks, where the system is regarded as a collection of elements and their pairwise relations. Network science has flourished over the last 15 years and currently plays a central role in the study of complexity.

In a network modelling approach, dynamical processes are typically incorporated as an additional ingredient on top of a static network structure. This scheme typically enables us to examine the interplay between structure and dynamics from two sides. On the one hand, one can assess how structural properties of a network influence a dynamical process, for example, how certain structural patterns change the speed of viral diffusion. On the other hand, one can explore the structure of a network by means of a dynamical process, for instance by identifying central nodes in terms of their impact on the dynamics.

By contrast, the study of temporal networks, the subject of this book, starts from the observation that real-world systems exhibit complex dynamical patterns that are not reproduced by models of dynamics defined on static networks. A proper description of temporal networks and their behaviour requires an integrated modelling of their structural and dynamical properties. In temporal networks, nodes and links are themselves dynamical entities. Their dynamics can be informed by empirical data and are themselves a main objective of modelling. Temporal networks are much more complex objects than static ones, as temporality radically alters even the most elementary concepts of network science, such as the path between two nodes, and their study requires additional computational and mathematical tools. Despite fairly recent interests in temporal networks, we have witnessed a convergence towards certain tools and models, as well as rapid accumulation of empirical observations of temporal networks in various domains.

The purpose of this monograph is to present traditional and modern techniques for temporal networks in a unified fashion. We have tried to incorporate methods scattered in different fields including statistical physics, computer science, statistics and applied mathematics. In fact, there already exist comprehensive review papers and an edited book on temporal networks. Our project differs from them in the following ways. First, we focus on tools, currently exploited and potentially used for studying temporal networks, rather than data themselves. Second, this book is not intended to be exhaustive. Rather, we have selected some sets of techniques to be explained in detail. We then proceeded in a pedagogical manner by introducing the reader to basic analysis tools before presenting advanced concepts. More generally, we believe that this book is a good entry point to the study of complex systems via an exposure to key modelling tools.

Acknowledgments

We first thank Henrik J. Jensen, who invited us to write the present book on the topic of our choice. The book owes to collaborations and discussion with numerous people as well as between the authors. Although it is impossible to mention all them, Naoki Masuda is especially indebted to his collaborators on projects closely related to this book: Tom Britton, Víctor M. Eguíluz, Byungnam Kahng, Jeon-Seop Kim, Konstantin Klemm, Hiroshi Kori, Jens Malmros, Shun Motegi, Mitsuhiro Nakamura, Ryosuke Nishi, Hisashi Ohtsuki, Kodai Saito, Nobuo Sato, Leo Speidel, Yuko K. Takahashi and Kazuo Yano. Likewise, Renaud Lambiotte is indebted to Mauricio Barahona, Vincent Blondel, Ed Bullmore, Timoteo Carletti, Jean-Charles Delvenne, Paul Expert, Cecilia Mascolo, Vsevolod Salnikov and Stefan Thurner for their inspiring and fruitful interactions over the years. He is also beholden to Radu Balescu, Léon Brenig, Michel Mareschal and Jean Wallenborn for carefully teaching him building blocks of science and revealing the beauty of the Boltzmann equation.

We are grateful to the following people for valuable feedback to the manuscript: Marya Bazzi, Sarah de Nigris, Tim Evans, Petter Holme, Vincenzo Nicosia, Mason A. Porter, Luis E. C. Rocha, Martin Rosvall, Jari Saramäki, Michael T. Schaub and Taro Takaguchi. We also acknowledge Alain Barrat for permitting us to use a figure generated from the website of the Sociopatterns project. Moreover, both of us warmly thank Sid Redner, who has played an important role in our respective careers, as an inspiration and collaborator. The title of this book is a tribute to his wonderful book on first-passage processes.

Naoki Masuda is indebted to his wife Chiho. Without her support this project would have been impossible. He also thanks his three daughters, Ami, Rika and Yuri for always cheering him up. Renaud Lambiotte warmly thanks Mélanie for bringing poetry into his life and Bram, Charlie and Thyl, who shorten his nights but make his days so much brighter.

Contents

Chapter 1

Introduction

A network is a collection of nodes and links, where a link connects two nodes. Many social, natural and engineered systems can be represented as networks. Examples include friendship networks, international relationships, gene regulatory networks, food webs, airport networks and the Internet just to name a few. Since the late 1990s, our understanding of real networks, from large to small ones, has been significantly advanced with the integration of theoretical, computational and conceptual tools from statistical physics, computer science, engineering, mathematics and other domains. Many networks have been recognised to be complex but governed by beautiful and universal laws. Together with applications, this field of research can be collectively called network science.

Evidence suggests that empirical networks observed in a variety of domains are far from static. It has been known since the advent of network science, and even earlier, that many networks grow in time. However, growth is not the only way in which networks evolve. Nodes and links may emerge and disappear during the lifetime of a network. For instance, the same link may be active just for a short period but repetitively with intermittent periods. Moreover, a rate of link activation may depend on time. Complex dynamical patterns of networks may arise for many reasons — circadian and weekly rhythms of actors, interactions between different links, stochasticity, finite response time of a system and so forth. These issues are at the core of the study of temporal networks, also called time/temporally varying networks/graphs, evolving graphs and evolutionary network analysis. The present book is about temporal networks and focuses on mathematical and computational tools to describe and model their behaviour.

As an illustration, let us consider the networks composed of pupils and teachers in a primary school at four time points in a day, shown in Fig. 1.1.

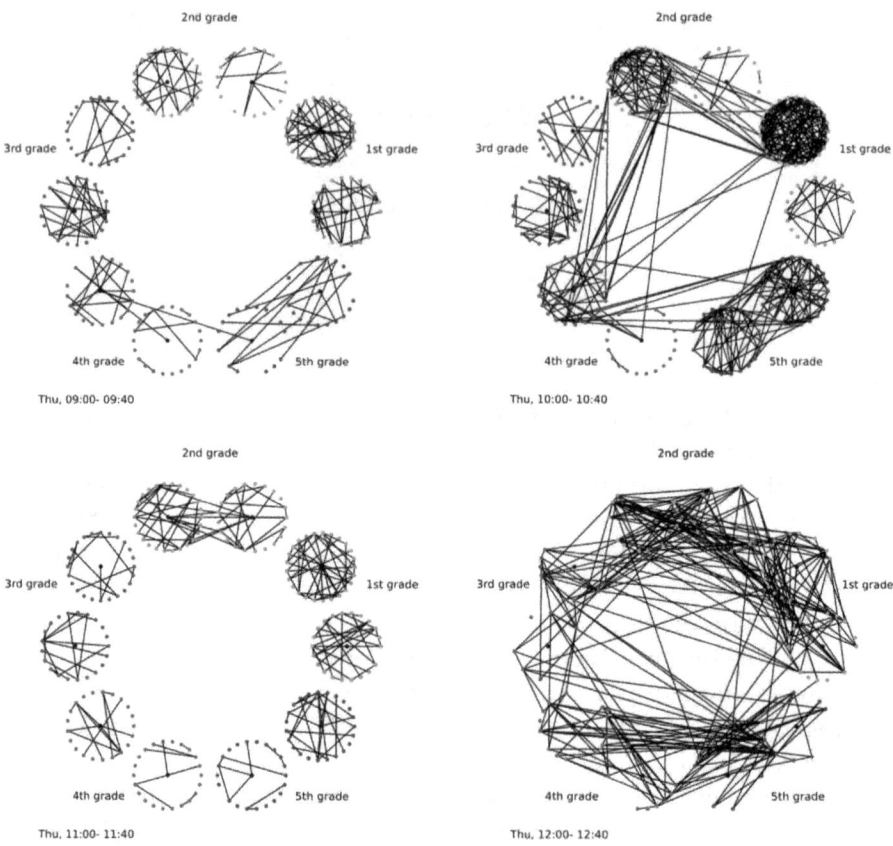

Fig. 1.1   Snapshots of dynamic face-to-face contact patterns in a primary school [Stehlé
et al. (2011)]. A circle represents a school class. A node in the centre of a circle is a
teacher. The other nodes are pupils. Networks at four time points are shown. This figure
is generated from the video available at http://www.sociopatterns.org/gallery/ with a
permission.

The figure indicates that the network varies over time, reflecting the daily
schedule of the school. We would lose a lot of information by aggregating
the four networks into one and discarding the temporal information in the
original data. This includes possible correlations between link activations,
the order in which links appear and disappear, circadian rhythms of human
activity and so on.

Why does temporality matter? Do we have to bother and consider
this additional dimension of networks on top of already complex static

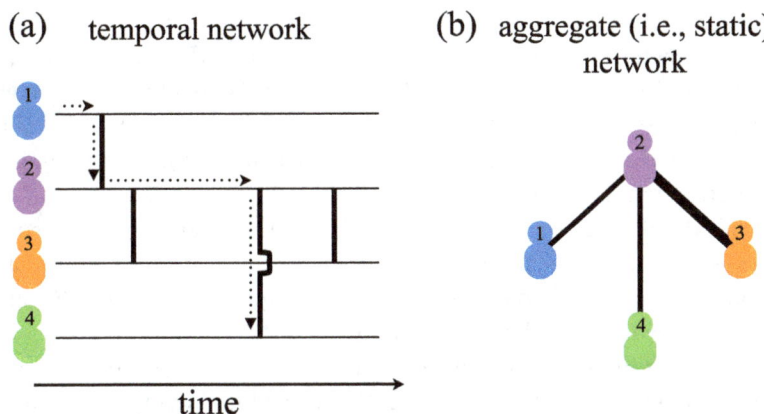

Fig. 1.2 Comparison of temporal and static networks. (a) A temporal network composed of four individuals and four events. (b) Aggregate network corresponding to the temporal network shown in (a). The width of the line represents the weight of the communication between the two individuals as quantified by the number of events observed on a link.

networks, making things even more complicated? Does the temporality of networks significantly alter our understanding of how networked systems function? Figure 1.2 provides affirmative evidence. Figure 1.2(a) shows a temporal network composed of four individuals. The time flows along the horizontal axis. Two individuals have a conversation event when there is a vertical line. There are four conversation events. In the third one, individuals 2 and 4 talk with each other without involving individual 3. If individual 1 starts propagating a news that the other three individuals have not heard of, the news reaches the entire network via conversation events. For example, individual 4 hears it from individual 2 through the path shown by the dotted arrows. However, if the news starts from individual 4, it cannot reach individual 1 within this temporal network. Although each conversation event is a symmetric event, allowing the information flow in both directions (e.g., from 1 to 2 and from 2 to 1), temporality of the network introduces asymmetry in the information flow at a network level.

If we aggregate the temporal network, we obtain the static network shown in Fig. 1.2(b), which we call the aggregate network. In the aggregate network, the information on the timing of the events is discarded, and the fact that individuals 2 and 3 communicate twice in Fig. 1.2(a) is represented by a link twice thicker than the other two links. Aggregate networks, which are static objects, are often deceptive. In Fig. 1.2(b), individual 2 connects to everybody else and acts as a hub. A news starting from anybody can eventually reach anybody else through individual 2. However, if we unfold the temporal information, we clearly see that this conclusion is incorrect (Fig. 1.2(a)). A news starting from individual 4 cannot reach individual 1. This example is anecdotal, but it warns us of the importance of temporality in understanding network phenomena.

In this book, we are interested in the combination of complexity of network structure and complexity induced by temporal behaviour of the network. There are already excellent review papers [Holme and Saramäki (2012); Aggarwal and Subbian (2014); Holme (2015)] and an edited book [Holme and Saramäki (2013)] on temporal networks. The objective of this book is to be an entry point to the mathematical and computational analysis of temporal networks. Therefore, we do not provide an exhaustive list of recent literature in the field. We instead provide an introduction to tools required for analysing temporal networks and focus on a selection of models and algorithms. Among the subjects that we do not treat in this book, let us mention adaptive networks, where the structure of the network and the state of the nodes co-evolve in time [Gross and Blasius (2008); Gross and Sayama (2009)]. In the following, we thus assume that temporal dynamics of networks are exogenously given by data or models.

The organisation of the book is as follows. In Chapter 2, we introduce mathematical tools from probability theory, linear algebra and others, which are used in the subsequent chapters. In Chapter 3, we summarise basics and a selection of advanced topics on temporally static networks. Because there are various monographs, review papers and online resources on static networks already (e.g., [Newman (2010); Barabási (2016)]), our exposure is not intended to be comprehensive but limited to the materials used in the later chapters on temporal networks. Therefore, we deliberately omitted some of the topics that would be contained in standard network science books. Examples include the Watts-Strogatz model, small-world effects, degree correlation, much of centrality measures, much of percolation, spatial networks and much of multilayer networks. Except for random walks, which are crucial in a variety of algorithms, dynamics on networks are not treated in this chapter. They are introduced later in Chapter 6. The last three chapters are the main chapters of this book, where we discuss temporal networks, with an emphasis on mathematical and computational tools. In Chapter 4, we explain various measurements and properties of temporal networks. In particular, we stand on two main representations of temporal networks among others and emphasise which representation is

Table 1.1   Main variables.

Variable	Meaning
N	number of nodes
M	number of links
$t_{\max}$	number of snapshots; sometimes the observation period
τ	inter-event time
$\psi(\tau)$	distribution of inter-event times
A	adjacency matrix
L	Laplacian matrix
d	distance, variously defined
Q	modularity
v_i	ith node
k_i	degree of node v_i
$\langle x \rangle$	mean of x

amenable to which measurements. In Chapter 5, we introduce generative models of temporal networks. In Chapter 6, we introduce a selection of models and analysis of dynamics on temporal networks.

Main notations used throughout the book are summarised in Table 1.1. Furthermore, we use two order notations; $O(f(x))$ in this book indicates that a quantity is proportional to $f(x)$ as $x \to \infty$. $o(f(x))$ indicates that a quantity is much less than $f(x)$ as $x \to \infty$. We use $\propto$ to indicate proportional to, and $\approx$ to approximately equal to.

We conclude this introductory chapter by stating some outstanding questions that warrant future work and could be tempting for the reader. First, many temporal networks occur as a result of mobility of individuals. Examples include social contact networks of humans, where a contact is defined by physical proximity, animals' networks due to animal trades and networks of robots exploring a two-dimensional arena. Relationships between mobility and temporal networks are yet unclear. In fact, mobility is regulated by spatial constraints. For example, individuals tend to visit nearby rather than remote sites. One place may not be able to host individuals beyond its capacity, which provides another constraint on mobility. Extending spatial networks [Barthélemy (2011)] and temporal networks to spatio-temporal networks may be an interesting line of research.

Second, related to spatial networks, metapopulation models (Sections 5.8 and 6.4.2) are a powerful framework with which to theoretically and empirically study the impact of human mobility on population dynamics, such as epidemic processes. Individuals in a metapopulation model network can be considered to form a temporal network. What is the relationship between the two?

Third, temporal networks are time series of networks. Time series analysis including multivariate ones has been studied for a long time in statistics, signal processing and other research communities. We will glimpse into the intersection between time series analysis and temporal networks when we discuss change point analysis (Section 4.10) and link prediction (Section 4.11). Nevertheless, various tools and concepts available in time series analysis have been underused for analysing temporal networks.

Fourth, related to time series analysis, there are methods to infer causality between time series, such as the Granger causality [Granger (1969)], transfer entropy [Schreiber (2000)] and convergent cross mapping [Sugihara *et al.* (2012)]. Once detected, causality may be represented as a directed link in a network. These methods as well as the concept of causality have been underemployed in the context of temporal networks (and perhaps static networks). Building frameworks to examine causality in temporal networks or using temporal network analysis tools to extract causality in multivariate data may be an interesting programme.

Finally, a crucial aspect of temporal networks is the coexistence of different time scales, i.e., one associated with the time evolution of the network, another associated with the speed at which dynamical processes take place on it, and also one associated with the temporal resolution of measurements. While different specific regimes have been considered in the literature, a truly united framework, integrating these different temporal processes, would constitute an important step towards our understanding of temporal networks.

Chapter 2

Mathematical toolbox

In this chapter, we introduce mathematical tools used in the subsequent chapters. They are a collection of elementary probability theory and linear algebra, and somewhat advanced materials on stochastic processes.

2.1 Probability

2.1.1 *Discrete variables*

A random variable is a variable that takes its value stochastically. A discrete random variable X is defined on a set S of possible values x such that $p(x) \geq 0$ for any $x \in S$ and $\sum_{x \in S} p(x) = 1$, where $p(x)$ is the probability that X takes value x; we use p to denote the probability throughout the book. A common example is a fair dice for which $S = \{1, 2, 3, 4, 5, 6\}$ and $p(x) = 1/6$ for each $x \in S$. If one throws the dice many times, the fraction of times with which one observes 1 tends to $1/6$.

A probabilistic event is specified by a certain subset of possible values in X. In the previous example, the event that a dice produces an odd number is represented as the event $X \in \{1, 3, 5\}$. When two events X_1 and X_2 are mutually exclusive, i.e., no value x belongs to both sets, we obtain

$$p(X_1 \text{ or } X_2) = p(X_1) + p(X_2). \tag{2.1}$$

Information about one event may inform the probability of another event. For instance, knowing that the dice produces an odd number increases the probability of $X = 1$ and decreases the probability of $X = 2$ to zero. Such information is quantified by the conditional probability. The conditional probability of X given Y is

$$p(X|Y) = \frac{p(X \text{ and } Y)}{p(Y)}. \tag{2.2}$$

By swapping X and Y in Eq. (2.2), we obtain $p(Y|X) = p(X \text{ and } Y)/p(X)$. By combining this equation with Eq. (2.2), we obtain the Bayes rule for conditional probabilities:

$$p(X|Y) = \frac{p(Y|X)p(X)}{p(Y)}. \tag{2.3}$$

Two events X and Y are said to be independent if the probability that X occurs is not affected by whether Y has occurred and vice versa. In other words,

$$p(X|Y) = p(X|\text{not } Y) = p(X). \tag{2.4}$$

When two events are independent, the probability that both events occur is the product of the probabilities that each event occurs, i.e.,

$$p(X \text{ and } Y) = p(X)p(Y). \tag{2.5}$$

In terms of the values of X and Y, we obtain

$$p(x, y) = p(x)p(y), \tag{2.6}$$

where $p(x, y)$ is the joint probability that $X = x$ and $Y = y$. The marginal distribution, i.e., the probability that $X = x$ regardless of the value of Y, is obtained by

$$p(x) = \sum_y p(x, y). \tag{2.7}$$

In principle, the random variable X can be either numerical (e.g., $1, 2, 3$) or non-numerical (e.g., white, red, black). In the former case, mostly relevant in this book, there exist different types of tools to characterise its properties. For instance, the expected value, or average, is defined as

$$\langle x \rangle = \sum_x xp(x). \tag{2.8}$$

We use $\langle \cdot \rangle$ to denote the mean throughout the book. The nth moment of X is defined by

$$\langle x^n \rangle = \sum_x x^n p(x), \tag{2.9}$$

where n is typically a positive integer, generalising Eq. (2.8). The second moment $\langle x^2 \rangle$ is related to the variance as follows:

$$\sigma^2 = \langle (x - \langle x \rangle)^2 \rangle = \langle x^2 \rangle - \langle x \rangle^2, \tag{2.10}$$

where σ is the standard deviation. Moments can be generalised to the case of multiple random variables, often to evaluate correlations between them. A familiar measure of linear dependence between two variables is the Pearson correlation coefficient defined by

$$\rho_{X,Y} = \frac{\langle (x - \langle x \rangle)(y - \langle y \rangle) \rangle}{\sigma_X \sigma_Y}, \tag{2.11}$$

where σ_X and σ_Y are the standard deviations of X and Y, respectively.

Here is a short list of frequently used discrete distributions:

(1) The Bernoulli distribution takes only two possible values, 0 or 1, i.e., failure or success, with probabilities $1 - p$ and p respectively. The mean $\langle x \rangle = p$ and the variance $\sigma^2 = p(1 - p)$.

(2) The binomial distribution describes the outcome of n independent and identically distributed random variables generated by the Bernoulli distribution with parameter p. The probability that exactly m successes are observed is given by

$$p(m) = \binom{n}{m} p^m (1 - p)^{n-m}, \tag{2.12}$$

where $0 \le m \le n$. Note that $p^m (1 - p)^{n-m}$ is the probability that a particular sequence containing exactly m successes is realised, and

$$\binom{n}{m} = \frac{n!}{m!(n - m)!} \tag{2.13}$$

is the number of sequences of length n that possess exactly m successes. We obtain $\langle m \rangle = np$ and $\sigma^2 = np(1 - p)$.

(3) The geometric distribution is defined via the waiting time before a success is observed, in a sequence of independent and identically distributed random variables obeying the Bernoulli distribution. The geometric distribution is defined as

$$p(m) = (1 - p)^m p, \tag{2.14}$$

where $m = 0, 1, \ldots$. The factor $(1 - p)^m$ corresponds to m consecutive failures, and p to the success on the $(m + 1)$th trial. We obtain $\langle m \rangle = (1 - p)/p$ and $\sigma^2 = (1 - p)/p^2$.

(4) The Poisson distribution is given as the limit of the binomial distribution as $n \to \infty$ while the mean np tends to a constant λ (therefore $p \to 0$). The Poisson distribution is given by

$$p(m) = \frac{m^\lambda e^{-\lambda}}{m!}, \tag{2.15}$$

where $m = 0, 1, \ldots$. We obtain $\langle m \rangle = \sigma^2 = \lambda$.

2.1.2 *Continuous variables*

Continuous random variables describe variables that take any value in a continuum of values, typically any real values or non-negative real values. Continuous random variables are set by their probability density function, $f(x)(\geq 0)$, defined such that the probability of observing any value between a and b is equal to

$$p(a \leq X \leq b) = \int_a^b f(x)\mathrm{d}x, \tag{2.16}$$

where $\int_{-\infty}^{\infty} f(x)\mathrm{d}x = 1$.

Most operations for discrete random variables are easily transferred to continuous random variables via replacement of sums by integrals. For instance, the joint probability density function $f(x, y)$ for continuous random variables satisfies

$$p(a \leq X \leq b, c \leq Y \leq d) = \int_a^b \int_c^d f(x, y)\mathrm{d}x\mathrm{d}y. \tag{2.17}$$

The moments of the distribution are given by

$$\langle x^n \rangle = \int_{-\infty}^{\infty} x^n f(x)\mathrm{d}x. \tag{2.18}$$

If two random variables are independent, their joint distribution factorises into the product of their marginals:

$$f(x, y) = f(x)f(y). \tag{2.19}$$

Finally, it is often practical to focus on the cumulative probability $F(x)$, defined as the probability that the variable takes a value smaller than x:

$$F(x) = \int_{-\infty}^{x} f(x')\mathrm{d}x'. \tag{2.20}$$

By definition, $F(-\infty) = 0$ and $F(\infty) = 1$. We also often use the complementary cumulative probability, also called the survival probability or survival function, given by

$$\tilde{F}(x) = \int_x^{\infty} f(x')\mathrm{d}x' = 1 - F(x). \tag{2.21}$$

Classical distributions for continuous random variables include the following ones:

(1) The uniform distribution takes a constant probability on interval $[a, b]$, i.e.,

$$f(x) = \frac{1}{b - a} \quad (a \leq x \leq b). \tag{2.22}$$

We obtain $\langle x \rangle = (b - a)/2$ and $\sigma^2 = (b - a)^2/12$.

(2) The exponential distribution is defined by

$$f(x) = \lambda e^{-\lambda x} \quad (x \geq 0). \tag{2.23}$$

Its cumulative distribution is given by

$$F(x) = 1 - e^{-\lambda x} \quad (x \geq 0). \tag{2.24}$$

We obtain $\langle x \rangle = 1/\lambda$ and $\sigma^2 = 1/\lambda^2$.

(3) The Gaussian or normal distribution is defined by

$$f(x) = \frac{1}{\sqrt{2\pi}\sigma} e^{-\frac{(x-\mu)^2}{2\sigma^2}} \quad (-\infty < x < \infty), \tag{2.25}$$

where μ is the average and σ^2 is the variance. The Gaussian distribution exhibits a bell shape. The Gaussian distribution can been seen as a continuous limit of the binomial distribution. The binomial distribution with n trials, each with probability p, converges to the Gaussian distribution with mean np and variance $np(1-p)$ owing to the central limit theorem. In particular for this reason, the Gaussian distribution is frequently observed in empirical data.

2.2 Renewal processes

2.2.1 *Poisson processes*

Let us consider a system where events take place in a discrete and apparently random fashion. Those events may be emails arriving in a mail box, or atoms colliding in a gas. Such systems are often modelled, as a first order approximation, by a Poisson process, also called the homogeneous Poisson process. The Poisson process assumes that the events are independent of each other, that the rate at which the events take place is constant over time and that time is continuous. These assumptions are often violated in empirical data. For instance, in the case of emails, their reception certainly depends on the time of the day and on the day of the week. In addition, emails are often not independent processes; an email may trigger a discussion thread between two users, causing a cascade of emails. These challenges are in fact a main focus of temporal network studies, as we will see in Chapters 4, 5 and 6. Yet, the Poisson processes are advantageous in their simplicity, which allows us to exactly calculate their properties and make them serve as a baseline model. Modelling of temporal networks often aims at understanding deviations from the Poisson process and relaxing

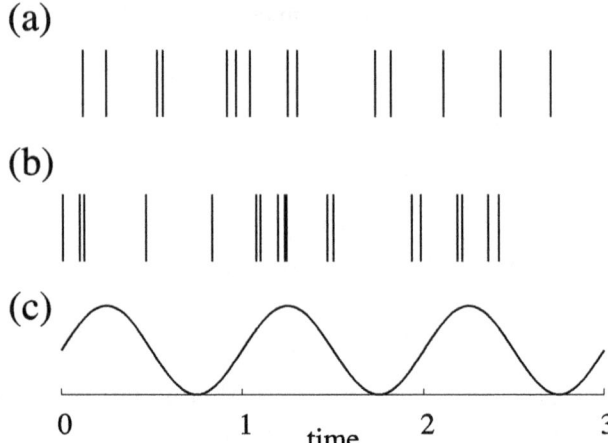

Fig. 2.1 Homogeneous and non-homogeneous Poisson processes. (a) An event sequence generated by the (homogeneous) Poisson process with $\lambda = 5$. (b) An event sequence generated by the non-homogeneous Poisson process with the sinusoidal rate shown in (c), i.e., $\lambda(t) = 5(1 + \sin 2\pi t)$.

the assumptions underlying the Poisson process to develop more realistic models.

The Poisson process is defined as follows. Consider a time window of duration Δt and the probability q that an event takes place within time Δt. By definition, the event rate is given by $\lambda = q/\Delta t$. A Poisson process is specified by the rate λ for infinitesimally small Δt. For λ to be well-defined, $q \to 0$ must be satisfied as $\Delta t \to 0$. Consistent with this requirement, we do not allow multiple events to occur in a time window when Δt is sufficiently small. An event sequence generated by a Poisson process is shown in Fig. 2.1(a).

Let us derive two key properties of Poisson processes:

(1) The distribution of inter-event times, i.e., time between consecutive events: Let $p(n, t)$ be the probability of observing n events in time window $[0, t]$. By definition, we obtain

$$
\begin{aligned}
q &= p(1, \Delta t) = \lambda \Delta t, \\
1 - q &= p(0, \Delta t) = 1 - \lambda \Delta t,
\end{aligned}
\tag{2.26}
$$

when Δt (and hence q) is small. For any $n \geq 1$, we obtain

$$
\begin{aligned}
p(n, t + \Delta t) &= p(n, t)p(0, \Delta t) + p(n - 1, t)p(1, \Delta t) \\
&= p(n, t)(1 - \lambda \Delta t) + p(n - 1, t)\lambda \Delta t.
\end{aligned}
\tag{2.27}
$$

Equation (2.27) holds true because, if there are n events in time window $[0, t + \Delta t]$, either there are n events in $[0, t]$ and no event in $[t, t + \Delta t]$, or there are $n - 1$ events in $[0, t]$ and one event in $[t, t + \Delta t]$. This equation relates the probability of a system at a certain time to that at a previous time, and hence is an example of master equation, which we will encounter many times in this book.

In the limit $\Delta t \to 0$, Eq. (2.27) is reduced to

$$\frac{\mathrm{d}p(n, t)}{\mathrm{d}t} = \lambda p(n - 1, t) - \lambda p(n, t). \tag{2.28}$$

For $n = 0$, we obtain

$$\frac{\mathrm{d}p(0, t)}{\mathrm{d}t} = -\lambda p(0, t), \tag{2.29}$$

which results in

$$p(0, t) = e^{-\lambda t}. \tag{2.30}$$

To derive Eq. (2.30), we have used the initial condition $p(0, 0) = 1$, i.e., no event has occurred at $t = 0$. Because $p(0, t)$ is equal to the probability that no event occurs in $[0, t]$, the probability that the first event occurs in $[t, t + \Delta t]$ is given by $p(0, t) - p(0, t + \Delta t)$. Equation (2.30) implies that the inter-event time between two consecutive events, denoted by τ, is distributed according to

$$\psi(\tau) = -\frac{\mathrm{d}p(0, \tau)}{\mathrm{d}\tau} = \lambda e^{-\lambda \tau}. \tag{2.31}$$

The inter-event time of a Poisson process is distributed according to the exponential distribution. The mean inter-event time is given by

$$\langle \tau \rangle = \int_0^\infty \tau \psi(\tau) \mathrm{d}\tau = \frac{1}{\lambda}. \tag{2.32}$$

In Poisson processes, different inter-event times τ are independent of each other because event times before the last event time t do not affect the time τ to the next event since t. This property is called the renewal property of a Poisson process. Poisson processes satisfy a stronger property, i.e., having no memory in the sense that

$$p(\tau > t_1 + t_2 | \tau > t_2) = p(\tau > t_1). \tag{2.33}$$

Equation (2.33) indicates that the length of time, t_2, for which we have waited, actually without an event, does not affect the time of the next event. The time to the next event starting from $t = t_2$, i.e., t_1, is independent of t_2 and obeys $\psi(t_1)$.

(2) The distribution of the number of events observed within a given time window: Using Eq. (2.28) recursively, we obtain

$$p(n,t) = \frac{(\lambda t)^n}{n!} e^{-\lambda t} \qquad (2.34)$$

for any $n \geq 0$. Therefore, the probability of observing n events in $[0, t]$ obeys the Poisson distribution with mean and variance equal to λt. As discussed in Section 2.1.1, the Poisson distribution is a limiting case of the binomial distribution when the number of trials is very large and the expected number of successes remains fixed. This interpretation is consistent with the discrete-time formulation of the Poisson process because in $[0, t]$, there are $t/\Delta t$ trials in each of which an event occurs with small probability q. Therefore, the number of events in $[0, t]$ is distributed according to the binomial distribution whose mean is equal to $(t/\Delta t) \times q = \lambda t$.

A method to generate an event sequence obeying a Poisson process is to generate events one by one by independently drawing the inter-event time τ according to Eq. (2.31). An alternative method when the final time $t_{\max}$ is specified is to first draw the number of events n in $[0, t_{\max}]$ according to the Poisson distribution with parameter $\lambda t_{\max}$. Then, distribute each of the n events independently and uniformly on $[0, t_{\max}]$. The second method exploits the memoryless property of Poisson processes.

Let us introduce two extensions of Poisson processes. The first is non-homogeneous (also called inhomogeneous) Poisson processes, in which the event rate $\lambda(t)$ is time-dependent. In other words, an event occurs in $[t, t + \Delta t]$ with probability $\lambda(t)\Delta t$. This model is motivated by the fact that event rates seem to vary over time in a majority of empirical data. An event sequence generated by a non-homogeneous Poisson process is shown in Fig. 2.1(b). In this example, the rate is modulated sinusoidally as shown in Fig. 2.1(c). For a non-homogeneous Poisson process, Eq. (2.34) is extended as

$$p(n,t) = \frac{\Lambda(t)^n}{n!} e^{-\Lambda(t)}, \qquad (2.35)$$

where

$$\Lambda(t) = \int_0^t \lambda(t') dt'. \qquad (2.36)$$

The distribution of inter-event times, conditioned by the last event at $t = 0$, is given by

$$\psi(\tau) = \lambda(\tau) e^{-\Lambda(\tau)}, \qquad (2.37)$$

which extends Eq. (2.31). It should be noted that Eq. (2.37) is properly normalised, i.e., $\int_0^\infty \psi(\tau) d\tau = 1$.

2.2.2 *General renewal processes*

The second extension of Poisson processes, called renewal processes, considers a general distribution of inter-event times, $\psi(\tau)$. The renewal property dictates that different inter-event times are independent of each other and drawn from the same distribution $\psi(\tau)$. When $\psi(\tau) = \lambda e^{-\lambda \tau}$, we recover a Poisson process. When $\psi(\tau) = \delta(\tau - 1)$, events periodically happen at all integer times.

To obtain the time of the nth event or the number of events in a given time period, we need to sum independent random variables generated according to $\psi(\tau)$. As we will see below, it is convenient to study the problem in a frequency domain. Because $\tau \geq 0$ is definite positive, consider the Laplace transform defined as

$$\hat{\psi}(s) = \int_0^\infty \psi(\tau) e^{-s\tau} \mathrm{d}\tau \equiv \langle e^{-s\tau} \rangle. \tag{2.38}$$

The Taylor expansion of Eq. (2.38), i.e.,

$$\hat{\psi}(s) = \sum_{n=0}^\infty (-1)^n \frac{\langle \tau^n \rangle s^n}{n!}, \tag{2.39}$$

implies that $\hat{\psi}(s)$ generates the moments of the original distribution whenever they exist. The inverse Laplace transform is given by an integration in the complex plane as follows:

$$\psi(\tau) = \frac{1}{2\pi i} \int_{c-i\infty}^{c+i\infty} \hat{\psi}(s) e^{s\tau} \mathrm{d}s, \tag{2.40}$$

where c is a real constant larger than the real part of all singularities of $\hat{\psi}(s)$.

By definition, $p(0, t)$ is the probability that no event has occurred up to time t and is given by

$$p(0, t) = \int_t^\infty \psi(t') \mathrm{d}t', \tag{2.41}$$

whose Laplace transform is equal to

$$\hat{p}(0, s) = \frac{1 - \hat{\psi}(s)}{s}. \tag{2.42}$$

The probability of observing one event in $[0, t]$ is equal to the probability that a first event occurs at a time t' before t and no subsequent event occurs between t' and t. Therefore, we obtain

$$p(1, t) = \int_0^t \psi(t') p(0, t - t') \mathrm{d}t'. \tag{2.43}$$

By using the fact that a convolution in time is translated into a product in the Laplace domain and applying Eq. (2.42) to the Laplace transform of Eq. (2.43), we obtain

$$\hat{p}(1, s) = \hat{\psi}(s) \frac{1 - \hat{\psi}(s)}{s}. \tag{2.44}$$

Similarly, because the probability density that n events occur at times t_1, t_2, ..., t_n and nowhere else in $[0, t]$ is given by $\psi(t_1)\psi(t_2 - t_1) \cdots \psi(t_n - t_{n-1})p(0, t - t_n)$, we obtain [Cox (1962); Grigolini *et al.* (2001)]

$$\hat{p}(n, s) = \left[\hat{\psi}(s)\right]^n \frac{1 - \hat{\psi}(s)}{s}. \tag{2.45}$$

Equation (2.45) relates two ways to count time: one in terms of the number of events, n, and the other in terms of the physical time, t.

To obtain the moments of $n(t)$, it is convenient to use the generating function defined by

$$G(t, z) = \sum_{n=0}^{\infty} p(n, t)z^n. \tag{2.46}$$

By differentiating Eq. (2.46) with respect to z once or twice and setting $z = 0$, we obtain $\langle n(t) \rangle = \partial G(t, z)/\partial z|_{z=0}$, which is called the renewal function, and $\langle n(t)(n(t) - 1) \rangle = \partial^2 G(t, z)/\partial z^2\big|_{z=0}$. With the use of Eq. (2.45), the generating function for n in the Laplace domain is given by

$$\begin{aligned}
\hat{G}(s, z) &= \sum_{n=0}^{\infty} \hat{p}(n, s)z^n \\
&= \sum_{n=0}^{\infty} \left[\hat{\psi}(s)\right]^n \frac{1 - \hat{\psi}(s)}{s} z^n \\
&= \frac{1 - \hat{\psi}(s)}{s\left[1 - \hat{\psi}(s)z\right]}.
\end{aligned} \tag{2.47}$$

Therefore, the mean number of events in the Laplace domain is given by

$$\langle \hat{n}(s) \rangle = \frac{\partial \hat{G}(s, z)}{\partial z}\bigg|_{z=1} = \frac{\hat{\psi}(s)}{s\left[1 - \hat{\psi}(s)\right]}. \tag{2.48}$$

The Laplace transform of the second moment is given by

$$\langle \hat{n^2}(s) \rangle = \frac{\partial^2 \hat{G}(s, z)}{\partial z^2}\bigg|_{z=1} + \langle \hat{n}(s) \rangle = \frac{\hat{\psi}(s)\left[1 + \hat{\psi}(s)\right]}{s\left[1 - \hat{\psi}(s)\right]^2}. \tag{2.49}$$

The asymptotic behaviour of $\langle n(t) \rangle$ as $t \to \infty$ is given by the limit of $\langle \hat{n}(s) \rangle$ as $s \to 0$. When the mean inter-event time is finite, we obtain

$$\hat{\psi}(s) = 1 - \langle \tau \rangle s + o(s) \tag{2.50}$$

for small s. Note that Eq. (2.50) is consistent with Eq. (2.39). Substitution of Eq. (2.50) in Eq. (2.48) yields

$$\langle \hat{n}(s) \rangle \approx \frac{1}{\langle \tau \rangle s^2}, \tag{2.51}$$

whose inverse Laplace transform gives

$$\langle n(t) \rangle = \frac{t}{\langle \tau \rangle}. \tag{2.52}$$

Equation (2.52) is called the elementary renewal theorem. The theorem indicates that the number of events grows linearly with time for sufficiently long times, irrespectively of the details of $\psi(\tau)$. However, the linear relation breaks down when $\langle \tau \rangle$ diverges (Section 4.6.5).

We recover Eq. (2.34) in the case of Poisson processes as follows. The Laplace transform for a Poisson process is given by

$$\hat{\psi}(s) = \int_0^\infty \lambda e^{-\lambda \tau} e^{-s\tau} \mathrm{d}\tau = \frac{\lambda}{s + \lambda}. \tag{2.53}$$

Substitution of Eq. (2.53) in Eq. (2.45) yields

$$\hat{p}(n, s) = \left(\frac{\lambda}{s + \lambda} \right)^n \frac{1}{s + \lambda}, \tag{2.54}$$

whose inverse Laplace transform is explicitly calculated as

$$p(n, t) = \frac{(\lambda t)^n}{n!} e^{-\lambda t}. \tag{2.55}$$

The Laplace transform is also useful for deriving the distribution of the time of the nth event. The probability density that n events occur at time $t_1, t_2, \ldots, t_n$ is given by $\psi(t_1)\psi(t_2 - t_1) \cdots \psi(t_n - t_{n-1})$. Therefore, the Laplace transform of the probability density function of the time of the nth event, t_n, is equal to $\left[\hat{\psi}(s) \right]^n$. For a Poisson process, we obtain $\left[\hat{\psi}(s) \right]^n = [\lambda/(s + \lambda)]^n$. The inverse Laplace transform gives

$$p(t_n = t) = \frac{\lambda^n}{(n - 1)!} t^{n-1} e^{-\lambda t}, \tag{2.56}$$

which is the Gamma distribution.

2.3   Random walks and diffusion

2.3.1   *Discrete time*

The Poisson processes provide a basic model for modelling temporal events, i.e., when random events take place. Random walk processes are its counterpart for modelling trajectories in space, i.e., when and where random events take place. Random walk processes are a standard tool to emulate diffusion on networks and also to extract information from the structure of networks, as we will show later. In this section, we derive some basic properties of random walk processes in their simplest setting, when they take place on a one-dimensional space (i.e., line) in discrete time.

In each discrete time step, a walker performs a jump whose length and direction are random variables. The probability density of transition is denoted by $f(r)$. In other words, the probability that the walker located at x arrives in the interval $[x + r, x + r + \Delta r]$ in one jump is equal to $f(r)\Delta r$. The normalisation condition is given by $\int_{-\infty}^{\infty} f(r)\mathrm{d}r = 1$.

Our aim is to derive the density of the probability density that the walker is located at x after t steps, denoted by $p(x; t)$. Under the assumption that jumps are independent events, we obtain the following master equation:

$$p(x; t) = \int_{-\infty}^{\infty} f(x - x')p(x'; t - 1)\mathrm{d}x' \qquad (2.57)$$

because the probability of visiting x at time t is the probability of having visited x' at time $t - 1$ and performing a jump of displacement $x - x'$.

Equation (2.57) for the entire range of x is more easily solved in the Fourier domain. The Fourier transform, analogous to the Laplace transform, is defined by

$$\hat{p}(k; t) \equiv \int_{-\infty}^{\infty} p(x; t)e^{-ikx}\mathrm{d}x, \qquad (2.58)$$

The original function is recovered through the inverse Fourier transform given by

$$p(x; t) = \frac{1}{2\pi} \int_{-\infty}^{\infty} \hat{p}(k; t)e^{ikx}\mathrm{d}k. \qquad (2.59)$$

Probability $p(x; t)$ is thus a combination of the oscillatory functions e^{ikx}, which form a base in the space of functions. The Fourier mode $\hat{p}(k; t)$ is the projection of $p(x; t)$ onto this base. The Fourier transform of $f(x)$, $\hat{f}(k)$, is called the structure function of the random walk. The Taylor expansion around $k = 0$ yields

$$\hat{p}(k;t) = \langle e^{-ikx} \rangle$$
$$= 1 - ik\langle x \rangle - \frac{1}{2}k^2\langle x^2 \rangle + O(k^3). \tag{2.60}$$

Equation (2.60) implies that the moments of $p(x;t)$ are obtained from the derivatives of $\hat{p}(k;t)$ at $k = 0$.

Like the Laplace transform, the Fourier transform transfers a convolution, such as Eq. (2.57), to a product. For this reason, working in the Fourier domain is often recommended when dealing with problems involving summations of random variables. Equation (2.57) is equivalent to

$$\hat{p}(k;t) = \hat{f}(k)\hat{p}(k;t-1). \tag{2.61}$$

If the walker is initially located at $x = 0$, such that $p(x;0) = \delta(x)$, which translates to $\hat{p}(k;0) = 1$, we obtain

$$\hat{p}(k;t) = \left[\hat{f}(k)\right]^t. \tag{2.62}$$

Using the inverse Fourier transform (Eq. (2.59)), the formal solution of the random walk in the time domain is given by

$$p(x;t) = \frac{1}{2\pi}\int_{-\infty}^{\infty} \left[\hat{f}(k)\right]^t e^{ikx} \mathrm{d}k. \tag{2.63}$$

This solution generally depends on the details of the structure function, $\hat{f}(k)$. However, the asymptotic behaviour of the random walk as t grows only depends on some of its properties. When the first two moments of the structure function are finite, the solution converges to the Gaussian profile

$$p(x;t) = \frac{1}{(2\pi Dt)^{1/2}} e^{-\frac{(x-vt)^2}{4Dt}} \tag{2.64}$$

with a variance growing linearly with time. More discussion on the topic, including its connection to the central limit theorem, stable and Lévy distributions, is provided in Appendix A.

2.3.2 *Continuous time*

We have looked at diffusion in terms of the number of steps. In this section, we consider continuous-time random walk models, which incorporate the timing of jumps into random walk processes. They are particularly useful for modelling diffusive dynamics on temporal networks in which the timing

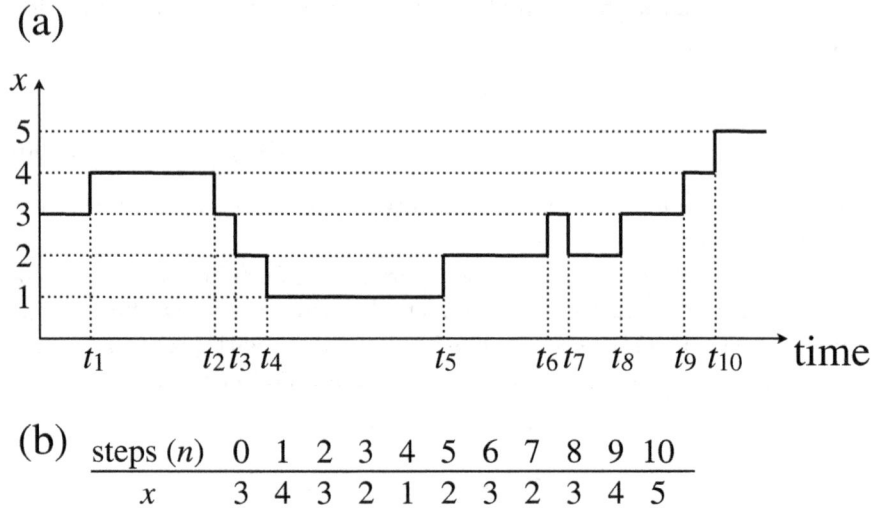

Fig. 2.2   Continuous-time random walk in the one-dimensional lattice. (a) Position of the walker, x, in real time, corresponding to $p(x; t)$. t_n represents the time of the nth jump. (b) Position of the walker after n jumps, corresponding to $p(x; n)$.

of events is a key feature (Section 6.3). The model is defined as follows. The walker waits between two jumps for a duration τ that independently obeys the probability density function $\psi(\tau)$, i.e., the jump events are generated by a renewal process. The time of the event and the length of the jump are assumed to be independent random events. A combination of $\psi(\tau)$ and the transition probability $f(r)$, where r is the displacement in a jump, determines the dynamical properties of the walker.

We denote by t_n the time of the nth jump event. By definition, $t_n = \sum_{i=1}^{n} \tau_i$, where each τ_i is independently and identically distributed according to $\psi(\tau)$. We learned the statistics of t_n, i.e., $p(n, t)$, in Section 2.2.2. Using $p(n, t)$, we derive the probability that the walker is located at x at time t, denoted by $p(x; t)$, in the following. It should be noted that $p(x; t)$ is distinguished from the probability that the walker is located at x after n steps, $p(x; n)$. The difference between $p(x; t)$ and $p(x; n)$ is illustrated in Fig. 2.2.

Because of the independence of the two processes, one generating r, the displacement of the jump, and the other generating τ, the inter-event time, we obtain

$$p(x;t) = \sum_{n=0}^{\infty} p(x;n)p(n,t). \tag{2.65}$$

Equation (2.65) reflects the fact that the walker may reach x at time t with various number of steps, n. By taking the Fourier and Laplace transform with respect to x and t, respectively, and using Eqs. (2.45) and (2.62), we obtain

$$\hat{p}(k;s) = \frac{1 - \hat{\psi}(s)}{s} \sum_{n=0}^{\infty} \hat{f}(k)^n \hat{\psi}(s)^n$$

$$= \frac{1 - \hat{\psi}(s)}{s} \frac{1}{1 - \hat{f}(k)\hat{\psi}(s)}. \tag{2.66}$$

This equation is the central result in the theory of continuous-time random walks. It will be generalised to the case of networks instead of the continuum space in Section 6.3. The inverse transform of Eq. (2.66) in terms of time and space provides $p(x;t)$. Behaviour of the random walk for large t is given by expanding $\hat{p}(k;s)$ or $\hat{p}(x;s)$ for small s, as in Section 2.2.2.

2.4 Power-law distributions

We have seen the emergence of two types of distributions in stochastic processes, the exponential distribution in the case of Poisson processes, and the Gaussian distribution in the case of random walk processes. Another type of distribution, i.e., power-law distribution, plays a central role in network science and in the theory of complex systems in general. In this section we overview properties of power-law distributions and raise some flags in order to properly use them when modelling complex systems. See [Newman (2005); Clauset *et al.* (2009)] for surveys.

We explain a power-law distribution for continuous variables, keeping in mind that most of the observations generalise to the case of discrete variables. Consider the Pareto distribution given by

$$p(x) = Cx^{-\alpha} \quad (x \geq x_{\min}), \tag{2.67}$$

where α is the power-law exponent of the distribution, $x_{\min}(> 0)$ is the minimum value taken by the random variable and $C = (\alpha - 1)x_{\min}^{\alpha-1}$ is the normalisation constant respecting

$$\int_{x_{\min}}^{\infty} Cx^{-\alpha}\mathrm{d}x = 1. \tag{2.68}$$

Other power-law distributions are by definition asymptotically (i.e., for large x) the same as Eq. (2.67) up to a normalisation constant.

Power-law distributions mainly differ from the exponential and Gaussian distributions by the significant mass of probability carried by their tail, i.e., large values of x. The exponential and Gaussian distributions have a characteristic scale such that the probability of observing instances many times larger than this scale is negligible. In contrast, under a power-law distribution, a vast majority of instances exhibits small values while few but non-negligible instances produce very large values. Power-law distributions are associated with a broad heterogeneity in the system and are said to have a fat or long tail, because the tail of the distribution is much more populated than in exponential-like distributions. Power-law distributions are typically found in the wealth of individuals, populations of cities, the frequency of words in text, sales of books and music, citations that a scientific paper receives and so forth. Since the advent of the Pareto distribution and the associated Zipf's law, power-law distributions have been studied over a century. We stress that fat tails are also present in distributions without a power-law tail. Examples include stretched exponential distributions and log-normal distributions.

The moments of power-law distributions are given by

$$\langle x^{\beta} \rangle = \int_{x_{\min}}^{\infty} x^{\beta} p(x) \mathrm{d}x = \frac{\alpha - 1}{\alpha - 1 - \beta} x_{\min}^{\beta} \quad (\beta < \alpha - 1). \qquad (2.69)$$

The moments for $\beta \geq \alpha - 1$ are divergent. In particular, the mean $\langle x \rangle$ does not exist for $1 < \alpha \leq 2$, and the variance does not exist for $2 < \alpha \leq 3$. These features impact various structural and dynamical properties of complex systems including networks, as we will see throughout this book. When $\alpha \leq 1$, the distribution is ill-defined because $\int_{x_{\min}}^{\infty} p(x) \mathrm{d}x$ is divergent such that $p(x)$ cannot be normalised. When a moment, $\langle x^{\beta} \rangle$, diverges, its empirical measurement diverges as the number of samples increases and $\langle x^{\beta} \rangle$ with β only slightly smaller than $\alpha - 1$ converges very slowly. Both the divergence and slow convergence of moments are due to the appearance of extreme values. For example, the sample mean for the power-law distribution with $\alpha = 2$ diverges as we accumulate samples.

In a majority of empirical data, the distribution can be close to Eq. (2.67) only in a certain range of the variable. However, key observations such as the divergent moments hold true as long as a distribution behaves the same as Eq. (2.67) when $x \to \infty$ up to a normalisation constant. For example, the Cauchy distribution given by $p(x) = 1/\left[\pi(1 + x^2)\right]$

is qualitatively the same as Eq. (2.67) with $\alpha = 2$ as $x \to \infty$. It should also be noted that the tail of an empirical distribution ceases to be a power-law beyond a certain scale because of the finiteness of the system. The finite size effect typically leads to exponential cut-offs. Therefore, the power-law regime, if present, usually dominates for values that are neither too small nor too large.

The heterogeneity of power-law distributions is often associated with the presence of inequalities in the system. What fraction w of the total wealth is held by a certain fraction of the richest people when the wealth distribution is given by Eq. (2.67)? To answer this question, let us first calculate the fraction of the people whose wealth is at least x_0:

$$p(x \geq x_0) = \int_{x_0}^{\infty} Cx^{-\alpha}dx = \left(\frac{x_0}{x_{\min}}\right)^{-\alpha+1}. \tag{2.70}$$

The fraction of wealth held by these richest people is given by

$$w(x_0) = \frac{\int_{x_0}^{\infty} x \cdot Cx^{-\alpha}dx}{\int_{x_{\min}}^{\infty} x \cdot Cx^{-\alpha}dx} = \left(\frac{x_0}{x_{\min}}\right)^{-\alpha+2} = [p(x \geq x_0)]^{\frac{\alpha-2}{\alpha-1}}, \tag{2.71}$$

where we have assumed that $\alpha > 2$ so that the average wealth is finite. Equation (2.71) neither depends on x_0 nor $x_{\min}$ explicitly, and it provides a direct relation between $w(x_0)$ and $p(x \geq x_0)$. This relation is often called the "80-20 rule", anecdotally meaning that 80% of the wealth is in the hands of the richest 20%. More precisely, setting $p(x \geq x_0) = 0.2$, $w(x_0) = 0.2^{(\alpha-2)/(\alpha-1)}$ can take any value between 0.2 and 1 depending on the value of α. In the limit $\alpha \to \infty$, the system does not exhibit a power-law tail, and we obtain $w(x_0) = 0.2$. In this case, the system is egalitarian. As α decreases, the tail of the distribution becomes fat and inequality grows. In the extreme situation with $\alpha \to 2$, the total wealth belongs to an infinitesimally small fraction of the richest people.

Other properties of power-law distributions include the following:

- Power-law distributions are scale-invariant because they satisfy

$$p(c_1 x) = c_2 p(x) \tag{2.72}$$

for large x, where c_1 and c_2 are constants. Equation (2.72) implies that multiplying the variable, or equivalently, changing the unit in which it is measured, does not affect properties of the system.
- Power-law distributions conveniently take the form of a straight line in a log-log plot because Eq. (2.67) is equivalent to

$$\log p(x) = \log C - \alpha x. \tag{2.73}$$

When testing if empirical data are power-law distributed, it is instructive (but not conclusive) to plot their distribution on the log-log scale.

Tauberian and Abelian types of theorems help us understand power-law tails of probability distributions and functions in general. They are particularly useful for analytically understanding the long-term behaviour of stochastic dynamics when power-law statistics come into play. In short, Tauberian and Abelian theorems are the inverse of each other. For the purpose of this book, the Tauberian theorem for the Laplace transform is stated as follows [Feller (1971); Klafter and Sokolov (2011)]:

Consider a function $f(t)$ whose asymptotic behaviour is given by $f(t) \approx t^{\rho-1}$ ($\rho > 0$) for large t. The Laplace transform (Eq. (2.38)) of $f(t)$ near $s = 0$ is given by $\hat{f}(s) \approx \Gamma(\rho)s^{-\rho}$, where $\Gamma(\rho)$ is the gamma function.

As we have seen, the long-time behaviour of a dynamical process is determined by its behaviour at small s in the Laplace domain. The Tauberian theorem is thus useful in estimating the asymptotic properties of a dynamical process. Here, we are interested in power-law distributions $p(x) \propto x^{-\alpha}$, where the power-law exponent $\alpha > 1$; otherwise, $p(x)$ would not be normalisable. However, the Tauberian theorem is applicable when the power-law exponent is less than one, i.e., $f(t) \approx t^{-(1-\rho)}$, where $\rho > 0$. To fill this gap, we must apply the Tauberian theorem to the integral of $p(x)$ of an appropriate order. As we integrate a power-law $p(x)$ once, the power-law exponent decreases by unity.

If $p(x) \approx Cx^{-\alpha}$ with $1 < \alpha < 2$ when x is large, we can apply the Tauberian theorem to the survival probability of $p(x)$, i.e.,

$$P(x) \equiv \int_x^\infty p(x')\mathrm{d}x' \approx \frac{Cx^{-\alpha+1}}{\alpha - 1} \tag{2.74}$$

with $\rho = 2 - \alpha$. The Laplace transform of the survival probability is given by

$$\hat{P}(s) = -C\Gamma(1 - \alpha)s^{\alpha-2}. \tag{2.75}$$

Note that $\Gamma(1 - \alpha) < 0$. We obtain the Laplace transform of the original distribution, $\hat{p}(s)$, through $\hat{p}(s) = 1 - s\hat{P}(s)$. When $2 < \alpha < 3$, we apply the Tauberian theorem to the integral of $P(x)$. By repeating the same argument, we obtain the expansion of $\hat{p}(s)$ around $s = 0$ as

$$\hat{p}(s) = \begin{cases} 1 + \Gamma(1 - \alpha)s^{\alpha-1} + o(s^{\alpha-1}) & (1 < \alpha < 2), \\ 1 - \langle x \rangle s + \Gamma(1 - \alpha)s^{\alpha-1} + o(s^{\alpha-1}) & (2 < \alpha < 3), \\ 1 - \langle x \rangle s + \frac{\langle x^2 \rangle}{2}s^2 + \Gamma(1 - \alpha)s^{\alpha-1} + o(s^{\alpha-1}) & (3 < \alpha < 4), \end{cases} \tag{2.76}$$

and similarly for $\alpha > 4$.

2.5   Maximum likelihood

The previous sections provide mechanisms by which certain families of distributions emerge. When we are confronted with empirical data, a crucial step is to find the parameter values that best reproduce the data, given a model. There exist different approaches to parameter fitting. The most popular one is probably the maximum likelihood method.

Consider a sequence of observations $\{x_i\}$ ($i = 1, 2, \ldots$). We are trying to fit a certain model whose parameter set is denoted by θ and is assumed to have a finite support for simplicity. Maximum likelihood dictates that the parameter values are chosen to maximise the probability with which the model generates the observed data. To this end, we calculate $p(\theta|\{x_i\})$, which is related to $p(\{x_i\}|\theta)$ by Bayes' law

$$p(\theta|\{x_i\}) = p(\{x_i\}|\theta)\frac{p(\theta)}{p(\{x_i\})}. \qquad (2.77)$$

By definition, the probability of observing certain data, $p(\{x_i\})$, is fixed, and it does affect the optimisation of θ. Moreover, in the absence of other information, it is convenient to assume that any values of θ are equally likely such that the prior distribution $p(\theta)$ is a constant. Then, $p(\theta|\{x_i\})$ and $p(\{x_i\}|\theta)$ are proportional to each other, and the locations of their maximum coincide. Therefore, it suffices to maximise $p(\theta|\{x_i\})$ in terms of θ.

As an example, consider the model in which each x_i independently obeys the same exponential distribution. The likelihood of the data is given by

$$\mathcal{L}(\{x_i\}|\lambda) = \prod_{i=1}^{n} p(x_i|\lambda), \qquad (2.78)$$

where $p(x|\lambda) = \lambda e^{-\lambda x}$ and n is the number of observations. To find the value of λ that maximises the likelihood, we conventionally maximise the logarithm of $\mathcal{L}$. The maximum of

$$\log \mathcal{L}(\{x_i\}|\lambda) = \log \prod_{i=1}^{n} p(x_i|\lambda) = n \log \lambda - \lambda \sum_{i=1}^{n} x_i \qquad (2.79)$$

is obtained via

$$\frac{\partial}{\partial \lambda} \log \mathcal{L}(\{x_i\}|\lambda) = 0, \qquad (2.80)$$

which leads to

$$\hat{\lambda} = \frac{1}{\frac{1}{n}\sum_{i=1}^{n} x_i} = \frac{1}{\langle x \rangle}. \qquad (2.81)$$

The maximum likelihood estimation is easy if the log likelihood takes an analytical form and its maximum is explicitly computed. Otherwise, we resort to numerical methods such as the expectation-maximisation algorithm.

There are also other situations in which likelihood maximisation needs to be done carefully. For example, suppose that we are fitting the power-law distribution, Eq. (2.67), to data ([Newman (2005); Clauset *et al.* (2009)]). Usually, a power-law distribution provides a good fit of empirical data in a regime excluding small x values. Therefore, we regard x_{min} as the point where the power-law regime starts, which we are interested in estimating in addition to the power-law exponent α. The log likelihood of the data under the power-law distribution is given by

$$\log \mathcal{L}(\{x_i\}|\alpha, x_{min}) = n \log\left(\frac{\alpha-1}{x_{min}}\right) - \alpha \sum_{i=1}^{n} \log\left(\frac{x_i}{x_{min}}\right). \qquad (2.82)$$

Setting $\partial \log \mathcal{L}/\partial \alpha = 0$ yields the maximum likelihood estimator given by

$$\hat{\alpha} = 1 + \frac{n}{\sum_{i=1}^{n} \log\left(\frac{x_i}{x_{min}}\right)}. \qquad (2.83)$$

However, finding the optimal x_{min} value is not a straightforward exercise because changing values of x_{min} also changes the number of observations, n, falling within the assumed power-law regime, i.e., $x \geq x_{min}$. The likelihood monotonically decreases with increasing n because the probability of observing an additional data point is always smaller than unity. Therefore, the maximum likelihood in terms of x_{min} is obtained by a trivial solution $\hat{x}_{min} = \max_i x_i$, yielding $n = 0$. Other techniques must be used to estimate $\hat{x}_{min}$. The minimisation of goodness-of-fit statistics, such as the Kolmogorov-Smirnov test, measuring the distance between the cumulative distribution of the empirical data and that of the model, is one such possibility [Clauset (2010)].

2.6   Entropy, information and similarity measures

The entropy of a random variable, denoted by H, is a measure of its uncertainty and quantifies how much we know about a variable before observing it. After the observation, we get rid of the uncertainty and thus gain information H about the system. For a discrete random variable X, entropy is defined as

$$H(X) = -\sum_{x} p(x) \log p(x). \qquad (2.84)$$

If X can take one of n states, we obtain $0 \leq H(X) \leq \log n$. The maximum value $H(X) = \log n$ is realised when $p(x)$ is the uniform density, i.e., when $p(x) = 1/n$ for all x. The minimum value $H(X) = 0$ is realised when X is deterministic, i.e., $p(x) = \delta_{x,x_0}$ for a specific x_0, where δ is Kronecker delta. In the latter case, we know the value of X before observing it, hence the lack of uncertainty.

The joint entropy $H(X,Y)$ of a pair of discrete random variables with joint distribution $p(x,y)$ is defined as

$$H(X,Y) = -\sum_x \sum_y p(x,y) \log p(x,y). \qquad (2.85)$$

The conditional entropy $H(Y|X)$ is defined as

$$H(Y|X) = -\sum_x \sum_y p(x,y) \log p(y|x)$$
$$= -\sum_x p(x)H(Y|X=x), \qquad (2.86)$$

and refers to the entropy of Y conditioned on the value of X and averaged over all possible values of X. The joint entropy and conditional entropy are related by the chain rule:

$$H(X,Y) = H(X) + H(Y|X). \qquad (2.87)$$

Equation (2.87) states that the total uncertainty about X and Y is simply the uncertainty about X, plus the average uncertainty about Y once X is known.

What does the knowledge of one variable tell us about another one? The conditional entropy $H(Y|X)$ addresses this question. More precisely, mutual information $I(X,Y)$ is defined as the amount of information gained on X by knowing the value of Y as follows:

$$I(X,Y) = H(X) - H(X|Y) = H(X) + H(Y) - H(X,Y). \qquad (2.88)$$

If Y is perfectly informative in the sense that it tells us everything about X, mutual information reduces to the entropy of X because $I(X,Y) = H(X) - H(X|Y) = H(X)$. Mutual information is rewritten as

$$I(X,Y) = \sum_x \sum_y p(x,y) \log \frac{p(y,x)}{p(x)p(y)}. \qquad (2.89)$$

Equations (2.88) and (2.89) show that mutual information is symmetric, i.e., $I(X,Y) = I(Y,X)$. Mutual information measures the cost of assuming that two variables are independent when they are in fact not. Mutual information captures non-linear correlations between random variables, in contrast to linear quantities such as the Pearson correlation coefficient.

In the analysis of temporal networks, we often have to compare networks at different times. Mutual information can serve to this end by specifying a distribution $p(x)$ that summarises a network. Other commonly used similarity measures include the Pearson correlation coefficient and the Jaccard index. The Pearson correlation for random variables is given by Eq. (2.11) and adapted for a list of pairwise observations $\{(x_i, y_i); 1 \leq i \leq n\}$ as follows:

$$\frac{\sum_{i=1}^{n}(x_i - \langle x \rangle)(y_i - \langle y \rangle)}{\sqrt{\sum_{i=1}^{n}(x_i - \langle x \rangle)^2 \sum_{i=1}^{n}(y_i - \langle y \rangle)^2}}, \tag{2.90}$$

where $\langle x \rangle = \sum_{i=1}^{n} x_i/n$ and $\langle y \rangle = \sum_{i=1}^{n} y_i/n$. The Jaccard index for two sets S_1 and S_2 is defined by

$$\frac{|S_1 \cap S_2|}{|S_1 \cup S_2|}, \tag{2.91}$$

where $|\cdot|$ denotes the number of elements in the set. The Jaccard index takes the largest value, 1, when $S_1 = S_2$. It takes the smallest value, 0, when S_1 and S_2 do not have any common element.

2.7 Matrix algebra

Matrices are a standard representation of networks including temporal networks. Properties of matrices are crucial in order to describe linear dynamical systems and at the core of several algorithms to extract structural information from networks. In this section, we provide a short, practical summary of results from linear algebra, emphasising what will be used in later chapters.

Consider an $N \times N$ matrix A. A vector $\boldsymbol{u}$ and scalar value λ satisfying

$$A\boldsymbol{u} = \lambda \boldsymbol{u} \tag{2.92}$$

are called the eigenvector and eigenvalue, respectively. There are at most N eigenvalues and associated eigenvectors. If A is a symmetric matrix, i.e., $A_{ij} = A_{ji}$ $(1 \leq i, j \leq N)$, all the eigenvalues λ_i $(1 \leq i \leq N)$ are real. In addition, the eigenvectors $\boldsymbol{u}_i$ associated with different eigenvalues λ_i are orthogonal, i.e., $\langle \boldsymbol{u}_i, \boldsymbol{u}_j \rangle = 0$ if $i \neq j$, where $\langle, \rangle$ is the inner product. Matrix A may have duplicated eigenvalues. Even in this case, we can select the set of N eigenvectors such that the orthogonality is respected.

Matrix A is decomposed as

$$A = \sum_{\ell=1}^{N} \lambda_\ell \boldsymbol{u}_\ell \boldsymbol{u}_\ell^\top, \tag{2.93}$$

where $\top$ represents the transposition. The validity of Eq. (2.93) is verified by multiplying an arbitrary eigenvector $\boldsymbol{u}_i$ to both sides of Eq. (2.93). Due to the orthogonality of the eigenvector, we obtain $A\boldsymbol{u}_i = \lambda_i \boldsymbol{u}_i$, assuming that the eigenvectors are properly normalised such that

$$\langle \boldsymbol{u}_\ell, \boldsymbol{u}_{\ell'} \rangle = \delta_{\ell\ell'}. \tag{2.94}$$

By combining Eqs. (2.93) and (2.94), we obtain

$$A^n = \sum_{\ell=1}^{N} \lambda_\ell^n \boldsymbol{u}_\ell \boldsymbol{u}_\ell^\top. \tag{2.95}$$

We are often interested in the extremal eigenvalue such as the largest eigenvalue of a symmetric matrix A, i.e., $\lambda_{\max}$. The Perron-Frobenius theorem guarantees that when all elements of A are strictly positive, $\lambda_{\max}$ is the isolated (i.e., not duplicated) largest eigenvalue. In addition, all elements of the corresponding eigenvector $\boldsymbol{u}_{\max}$, called the Perron-Frobenius vector, have the same sign. Any other eigenvector $\boldsymbol{u}_\ell$ does not show this property because, due to the orthogonality $\langle \boldsymbol{u}_\ell, \boldsymbol{u}_{\max} \rangle = 0$, some of the elements in $\boldsymbol{u}_\ell$ must have the opposite signs. The Perron-Frobenius theorem also holds true for asymmetric matrices. In the asymmetric case, the statement that the largest eigenvalue is isolated is replaced by that of the modulus, or the absolute value of the eigenvalue. Matrices appearing in network analysis are often sparse, with a majority of elements being zero. The Perron-Frobenius theorem is also applicable in this situation if matrix A is primitive, i.e., if all elements of A are non-negative and all elements of A^n are positive for some integer $n > 0$. If an undirected network of interest is connected as a single component, which is usually the case in theoretical studies, matrices representing the network are usually primitive (with the exception of so-called bipartite graphs), such that the Perron-Frobenius theorem can be used.

The power method is a computationally efficient method to calculate $\lambda_{\max}$ and $\boldsymbol{u}_{\max}$ of a given matrix. To do this, we start with an (almost) arbitrary initial vector $\boldsymbol{x}$ and repeat multiplying A. By multiplying $\boldsymbol{x}$ to both sides of Eq. (2.93), we obtain

$$\boldsymbol{x}(1) \equiv A\boldsymbol{x} = \sum_{\ell=1}^{N} \lambda_\ell \boldsymbol{u}_\ell \langle \boldsymbol{u}_\ell^\top, \boldsymbol{x} \rangle \tag{2.96}$$

By repeating the multiplication of A on both sides of Eq. (2.96), we obtain

$$\boldsymbol{x}(n) \equiv A^n \boldsymbol{x} = A^n \boldsymbol{x}(n-1) = \sum_{\ell=1}^{N} \lambda_\ell^n \boldsymbol{u}_\ell \langle \boldsymbol{u}_\ell^\top, \boldsymbol{x} \rangle. \tag{2.97}$$

If $\lambda_{\max}$ is the isolated eigenvalue, as in the case of the primitive matrix, $\lambda_{\max}^n \gg \lambda_\ell^n$ for any other eigenvalue λ_ℓ for large n. Then, in Eq. (2.97), all but the one term corresponding to $\lambda_{\max}$ is negligible on the right-hand side as $n \to \infty$. After many iterations, we can obtain the largest eigenvalue $\lambda_{\max}$ by looking at how much each element of $\boldsymbol{x}(n)$ grows by one iterate and the corresponding eigenvector $\boldsymbol{u}_{\max}$ from $\boldsymbol{x}(n)$. In practice, we normalise $\boldsymbol{x}(n)$ in each iterate to avoid the elements of $\boldsymbol{x}(n)$ to become very large or small.

2.8 Linear stability

Network phenomena are often interpreted as dynamics placed on nodes that are coupled via links, providing a coupled dynamical system. Examples include interacting oscillators in fireflies and dynamics of mood in social systems. The most basic configuration of such a coupled dynamical system is an equilibrium of the dynamics, which is stable or perhaps unstable against perturbation. In this section, we explain a simple and powerful method, called linear stability analysis, to determine the stability of equilibria of coupled and generally non-linear dynamical systems.

Consider a set of N dynamical variables x_i ($1 \le i \le N$) evolving due to some internal dynamics and pairwise coupling with other elements. The dynamics are described by the following set of non-linear differential equations:

$$\frac{\mathrm{d}x_i}{\mathrm{d}t} = f_i(x_i) + \sum_{j=1}^{N} g_{ij}(x_i, x_j) \quad (1 \le i \le N), \tag{2.98}$$

where $f_i(x_i)$ and $g_{ij}(x_i, x_j)$ represent the intrinsic dynamics and the interactions between the variables, respectively. Let us assume that the dynamics have an equilibrium, i.e., stationary solution, x_i^* ($1 \le i \le N$) such that

$$f_i(x_i^*) + \sum_{j=1}^{N} g_{ij}(x_i^*, x_j^*) = 0. \tag{2.99}$$

The stability of the equilibrium is determined by considering a small deviation around it parameterised as $x_i = x_i^* + \epsilon_i$ ($1 \le i \le N$). By performing the Taylor expansion of Eq. (2.98) in terms of ϵ_i and only retaining the first-order terms, we obtain the linearised dynamics given by

$$\frac{\mathrm{d}\epsilon_i}{\mathrm{d}t} = f_i(x_i^*) + \frac{\partial f_i(x_i^*)}{\partial x_i}\epsilon_i$$

$$+ \sum_{j=1}^{N}\left[g_{ij}(x_i^*,x_j^*) + \frac{\partial g_{ij}(x_i^*,x_j^*)}{\partial x_i}\epsilon_i + \frac{\partial g_{ij}(x_i^*,x_j^*)}{\partial x_j}\epsilon_j\right]$$

$$= \sum_{j=1}^{N} J_{ij}\epsilon_j, \qquad (2.100)$$

where

$$J_{ij} = \delta_{ij}\left[\frac{\partial f_i(x_i^*)}{\partial x_i} + \sum_{\ell=1}^{N}\frac{\partial g_{i\ell}(x_i^*,x_\ell^*)}{\partial x_i}\right] + \frac{\partial g_{ij}(x_i^*,x_j^*)}{\partial x_j}. \qquad (2.101)$$

To derive the second line in Eq. (2.100), we used Eq. (2.99).

If J is diagonalised by the transformation $\Lambda = Q^{-1}JQ$, where Q is a non-degenerate matrix, the diagonal elements of Λ are the eigenvalues of J, which we denote by λ_i ($1 \le i \le N$). By solving Eq. (2.100) for small t, where the perturbation remains small, we obtain

$$\epsilon = e^{Jt}\epsilon(t=0) = Qe^{\Lambda t}Q^{-1}\epsilon(t=0), \qquad (2.102)$$

where $\epsilon = (\epsilon_1,\ldots,\epsilon_N)^\top$ is the time-dependent perturbation and would explode if an eigenvalue of J has a positive real part. If all the eigenvalues have a negative real part, ϵ would instead decay to zero. Therefore, the stability of the equilibrium is determined by the eigenvalues of J. Note that J differs for different equilibria even in the same dynamical system, Eq. (2.98).

2.9 Markov chains

Markov chains are stochastic dynamics on N states in discrete time. A state may be the position in a network having N nodes such that the process represents a random walk on the network. Alternatively, a state may be the number of infected people, between 0 and $N-1$, in a structureless population of $N-1$ individuals. In both cases, we number the states as 1, 2, ..., N. The state at time t ($t = 0, 1, \ldots$), which is a random variable, is denoted by X_t.

In a stochastic process on N states in general, state X_{t+1} may depend on all preceding states (i.e., full history) of the dynamics, i.e., X_0, X_1, ..., X_t. Under the Markov assumption, the conditional probability to observe a state at time $t + 1$ only depends on the state at time t. In other words, a discrete-time stochastic process verifying

$$p(X_{t+1} = i_{t+1} | X_t = i_t, \ldots, X_1 = i_1, X_0 = i_0) = p(X_{t+1} = i_{t+1} | X_t = i_t), \tag{2.103}$$

is called the Markov chain. Among the class of Markov chains, we are often interested in the stationary ones, in which the conditional state-transition probability does not depend on t:

$$p(X_{t+1} = j | X_t = i) \equiv T_{ij}. \tag{2.104}$$

Processes verifying both properties, Markovianity and stationarity, are called stationary Markov chains. Because a realisation of the process visiting state i must go somewhere including itself in the next time step, we obtain

$$\sum_{j=1}^{N} T_{ij} = 1. \tag{2.105}$$

A stationary Markov chain is fully described by an initial state and an $N \times N$ transition matrix $T = (T_{ij})$. The probability that state i is visited at time t, denoted by $p_i(t)$, evolves according to

$$p_j(t+1) = \sum_{i=1}^{N} p_i(t) T_{ij} \quad (1 \le j \le N). \tag{2.106}$$

It should be noted that $\sum_{i=1}^{N} p_i(t) = 1$ for any t, if the initial condition is properly normalised. Equation (2.106) is compactly rewritten as

$$\boldsymbol{p}(t+1) = \boldsymbol{p}(t) T, \tag{2.107}$$

where $\boldsymbol{p}(t) = (p_1(t) \cdots p_N(t))$. Equation (2.107) yields

$$\boldsymbol{p}(t) = \boldsymbol{p}(0) T^t. \tag{2.108}$$

A Markov chain is composed of different types of states. By definition, the process does not escape from an absorbing state once it has been reached. State i is absorbing if and only if $T_{ii} = 1$, which implies that $T_{ij} = 0$ for any $j \ne i$. A group of states forms an ergodic set if it is possible to go from i to j for any states i and j in the set and if the process does not leave the set once the process has reached it. An absorbing state is thus an ergodic set composed of a single state. Finally, a state is called a transient state if it is not a member of an ergodic set.

We denote the stationary density by $\boldsymbol{p}^* = (p_1^*, \ldots, p_N^*)$, where $p_i^* = \lim_{t \to \infty} p_i(t)$ $(1 \le i \le N)$ and hence $\sum_{i=1}^{N} p_i^* = 1$. Substitution of $p_i(t) =$

$p_i(t + 1) = p_i^*$ $(1 \leq i \leq N)$, which holds true in the limit $t \to \infty$, in Eq. (2.106) yields

$$\boldsymbol{p}^* = \boldsymbol{p}^* T. \tag{2.109}$$

Therefore, the stationary density is the left eigenvector of T with eigenvalue unity. Because

$$T \begin{pmatrix} 1 \\ \vdots \\ 1 \end{pmatrix} = \begin{pmatrix} 1 \\ \vdots \\ 1 \end{pmatrix}, \tag{2.110}$$

which is a consequence of Eq. (2.105), T is guaranteed to have an eigenvalue of unity. If the entire set of the N states is ergodic, one can go from i to j for any i and j. In this case, $\boldsymbol{p}^*$ is unique, and iterates of Eq. (2.108) starting from an almost arbitrary initial condition converge $\boldsymbol{p}^*$ except in special cases.

Then, the eigenvalue of unity is in fact the largest eigenvalue of T in terms of the modulus (i.e., absolute value). Therefore, $\boldsymbol{p}^*$ is the Perron-Frobenius vector. This observation is consistent with the fact that all elements of the Perron-Frobenius vector are positive (Section 2.7). In addition, Eq. (2.95) adapted to the case of asymmetric matrices dictates that the discrepancy of $\boldsymbol{p}(t)$ from $\boldsymbol{p}^*$ decays exponentially as $\propto |\lambda_{2nd}|^t$, where λ_{2nd} is the second largest eigenvalue of T in terms of the modulus. In words, the second largest eigenvalue governs the relaxation time of the iterate. More generally, the speed of convergence is determined by the difference or ratio between λ_{2nd} and λ_{max}, with the latter being equal to unity in the current case. Therefore, we often call $1 - \lambda_{2nd}$ the spectral gap. A Markov chain with a large spectral gap converges rapidly.

Markov chain theory also allows us to answer other types of questions. For example, how long on average do the dynamics need to reach a certain state? What is the probability of ending in a certain absorbing state, depending on the initial condition? A short summary on these topics is provided in Appendix B.

2.10 Branching processes

A branching process is a Markov process in which each individual produces some (possibly zero) individuals and then dies, each of the new individuals undergoes reproduction, and so forth (Fig. 2.3). In network theory,

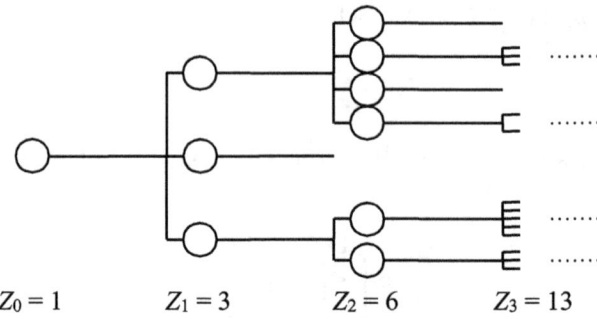

$$Z_0 = 1 \qquad Z_1 = 3 \qquad Z_2 = 6 \qquad Z_3 = 13$$

Fig. 2.3   Schematic of the Galton-Watson branching process.

branching processes are a useful tool for understanding network generation
and epidemic processes on networks. In network generation, we start from
a given node and explore its neighbours, neighbours of neighbours, and so
on to expand the network under investigation. In epidemic processes, an
initially infected node typically propagates infection to a certain number of
neighbouring nodes, each of which then infects some others, and so on. In
both cases, the number of nodes that a node newly recruits usually depends
on the node and hence can be considered as a random number, as assumed
in branching processes.

The Galton-Watson process is a prototypical branching process model
defined as follows. Fix the distribution of the number of offspring, $\{p(n)\}$,
where $p(n)$ is the probability that an individual reproduces n individuals.
The number of individuals in the tth generation is denoted by Z_t (Fig. 2.3).
First, there is initially one individual, i.e., $Z_0 = 1$. Second, this individual
generates offspring whose number Z_1 is drawn from $\{p(n)\}$. Third, each
of the Z_1 individuals in the first generation produces offspring whose number independently obeys distribution $\{p(n)\}$. The individuals born in this
stage, which total Z_2, define the second generation. We repeat this procedure to define further generations until the process gets extinguished. The
extinction may not occur, in which case the number of individuals grows
indefinitely.

The extinction requires $p(0) > 0$. In other words, an individual does
not produce any offspring with a positive probability. If $p(n)$ for large n
values is large, the population would grow rather than shrink. In fact, the
mean number of offspring, i.e., $\langle n \rangle \equiv \sum_{n=0}^{\infty} np(n)$ is the main determinant
of a branching process. If $\langle n \rangle \leq 1$, a realisation of the process will always

die out for sufficiently large t, except in the deterministic case $n = \langle n \rangle = 1$ such that each individual always yields exactly one offspring. In particular, $E[Z_t] = \langle n \rangle^t \to 0$ as $t \to \infty$. If $\langle n \rangle > 1$, $E[Z_t]$ exponentially grows and individual realisations of the process may grow as well.

We denote by q the probability that the process starting from one individual eventually dies out and, as we now show, $q = 1$ when $\langle n \rangle \leq 1$. If an individual produces n individuals, then the process will die out with probability q^n because of the independence of the sub-processes starting from n individuals. Therefore, we obtain the recursive relationship

$$q = \sum_{n=0}^{\infty} p(n)q^n. \tag{2.111}$$

Equation (2.111) always has $q = 1$ as a solution. It has a solution with $q < 1$ if and only if $\langle n \rangle > 1$. To show this, we use the fact that the solution is the intersection of $y = f_1(q) \equiv q$ and $y = f_2(q) \equiv \sum_{n=0}^{\infty} p(n)q^n$. Because $\langle n \rangle > 1$, it suffices to consider the case $p(0) + p(1) < 1$. If $p(0) = 0$, $q = 0$ is a solution because $f_2(0) = p(0) = 0 = f_1(0)$. If $p(0) > 0$, we obtain $0 < f_2(0) < 1$. Because $f_1(1) = f_2(1) = 1$, and $df_2(q)/dq = \sum_{n=1}^{\infty} np(n)q^{n-1} > 0$ and $d^2 f_2(q)/dq^2 = \sum_{n=2}^{\infty} n(n-1)p(n)q^{n-2} > 0$ when $0 < q \leq 1$, $y = f_1(q)$ and $y = f_2(q)$ cross in $0 < q \leq 1$ if and only if $df_2(q)/dq > 1$ at $q = 1$. This condition is equivalent to $\langle n \rangle > 1$. In this case, the process grows exponentially with probability $1 - q$.

Chapter 3

Static networks

In this chapter, we give a relatively short summary of static networks. We focus on what we will use in the subsequent chapters on temporal networks. For more extended reviews, see the literature mentioned in Chapter 1.

3.1 Definition

A network is a system made of nodes connected by links. Links can be undirected or directed, and unweighted or weighted. In the mathematical literature, a network is called a graph. It is defined as

$$\mathcal{G} = (V, E), \tag{3.1}$$

where V is a set of nodes (also called vertices) and E is a set of links (also called edges). The number of nodes and that of links are denoted by N and M throughout this book. Each link is defined by a pair of nodes, i.e., $e = (v, v') \in E$. In the case of undirected networks, the order of v and v' does not matter. In the case of directed networks, (v, v') indicates a link from v to v', and if $(v, v') \in E$ and $(v', v) \in E$, the two nodes are reciprocally connected. In the case of weighted networks, links are also assigned with a weight function, characterising the importance or weight of the link. An undirected and unweighted network with $N = 5$ nodes and $M = 6$ links is shown in Fig. 3.1.

In order to efficiently store networks and to carry out computations, it is necessary to use appropriate data structure [Skiena (2008)]. Each representation emphasises a certain aspect of the network and is amenable to certain types of computational or mathematical operations. We introduce two major representations.

A network can be represented by the corresponding $N \times N$ adjacency

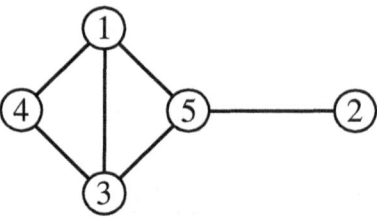

Fig. 3.1   An undirected and unweighted network.

matrix. Being adjacent means that two nodes are directly connected by a link. In the case of unweighted networks, the entries of the adjacency matrix are given by

$$A_{ij} = \begin{cases} 1 & \text{if node } v_i \text{ is adjacent to node } v_j, \\ 0 & \text{otherwise.} \end{cases} \tag{3.2}$$

If the network is weighted, A_{ij} can take positive values different from one, representing the weight of the link. In general, undirected and directed networks will yield symmetric and asymmetric adjacency matrices, respectively. The adjacency matrix of the network illustrated in Fig. 3.1 is given by

$$\begin{pmatrix} 0 & 0 & 1 & 1 & 1 \\ 0 & 0 & 0 & 0 & 1 \\ 1 & 0 & 0 & 1 & 1 \\ 1 & 0 & 1 & 0 & 0 \\ 1 & 1 & 1 & 0 & 0 \end{pmatrix}. \tag{3.3}$$

An adjacency matrix representation is useful for formulating and theoretically analysing structure of, and dynamical processes on networks. In particular, it is amenable to tools from linear algebra such as the analysis of eigenvectors and eigenvalues. A drawback of this representation is its memory cost because a network possessing N nodes requires $O(N^2)$ elements for storage. A majority of networks found in the real world and generated from models are sparse such that most elements of the adjacency matrix are equal to zero. Therefore, it is often preferable to use the data structure called sparse matrix. Its advantage is a significant gain in memory and faster computations because operations involving zeros are not executed.

In fact, a matrix formulation is often unsuitable even if sparse matrix representations are employed. This is the case when, for example, the computation of shortest paths is involved or all links have to be scanned repeatedly. A representation of static networks alternative to the adjacency matrix is called the link list. In this link-centric approach, a graph is described as a list of pairs of nodes, each corresponding to a link in the network, as follows:

$$\{(u_1, v_1), (u_2, v_2), \ldots, (u_M, v_M)\}. \tag{3.4}$$

When the network is directed, we interpret Eq. (3.4) as representing directed links from u_i to v_i $(1 \leq i \leq M)$. The link list of the network shown in Fig. 3.1 is given by

$$\{(v_1, v_3), (v_1, v_4), (v_1, v_5), (v_2, v_5), (v_3, v_4), (v_3, v_5)\}. \tag{3.5}$$

Link lists have the additional advantage of being efficiently used for link randomisation and numerical simulations of dynamics on sparse networks.

3.2 Degree distribution

The degree is defined as the number of links incident to a node. We denote the degree of the ith node by k_i. For undirected networks, the degree is given by

$$k_i = \sum_{j=1}^{N} A_{ij} \left(= \sum_{j=1}^{N} A_{ji} \right). \tag{3.6}$$

A network is called regular if all nodes have the same degree, i.e., $k_i = k_j$ for all i and j.

For directed networks, we distinguish the in-degree, i.e., the number of links incoming to the node, and the out-degree, i.e., the number of links outgoing from the node. They are given by $k_i^{\text{in}} = \sum_{j=1}^{N} A_{ji}$ and $k_i^{\text{out}} = \sum_{j=1}^{N} A_{ij}$, respectively.

Each link has two endpoints and hence contributes to the degree of two nodes by one each. Therefore, we obtain

$$\sum_{i=1}^{N} k_i = \sum_{i=1}^{N} \sum_{j=1}^{N} A_{ij} = 2M \tag{3.7}$$

for undirected networks. Equation (3.7) is called the handshaking lemma. It implies that the sum of the degrees of all the nodes in any undirected network is an even number. For directed networks, the handshaking lemma is given by $\sum_{i=1}^{N} k_i^{\text{in}} = \sum_{i=1}^{N} k_i^{\text{out}} = M$.

The degree distribution of a network is the frequency distribution of the degree and denoted by $p(k)$. A majority of networks in different domains possesses long-tailed degree distributions. In many situations, their tail is described by a power-law, i.e.,

$$p(k) \propto k^{-\gamma}, \tag{3.8}$$

where γ is typically between two and three. Because the maximum degree is equal to $N - 1$, Eq. (3.8) approximately holds true up to a certain cut-off degree, above which $p(k)$ rapidly decays to zero. The average degree, denoted by $\langle k \rangle$, is given by

$$\langle k \rangle = \sum_k k p(k). \tag{3.9}$$

The friendship paradox is a phenomenon, in which, anecdotally, the average number of friends of a friend is greater than the average number of friends of an individual [Feld (1991)]. This is purely a mathematical consequence that always arises unless every node has the same degree. The paradox originates from the fact that nodes with a large degree contribute disproportionately to the average degree of a friend, as they have a higher probability of being friends than low degree nodes do. Consider the situation shown in Fig. 3.2, where we pretend not to know the actual connection between nodes. This is in fact the definition of the configuration network model (Section 3.8.2). The network has $N = 6$ nodes and the average degree of a randomly selected individual is equal to

$$\frac{1}{6}(1 + 2 + 3 + 1 + 4 + 1) \equiv \langle k \rangle = 2. \tag{3.10}$$

To calculate the average number of friends of a friend, we have instead to perform a weighted average, accounting for the fact that a node with degree k will appear k times in the calculation of the average. The weighted average degree is equal to

$$\frac{(1 \times 1 + 2 \times 2 + 3 \times 3 + 1 \times 1 + 4 \times 4 + 1 \times 1)}{(1 + 2 + 3 + 1 + 4 + 1)} \approx 2.67, \tag{3.11}$$

where we have allowed self-loops and multiple edges for simplicity. In general, for sufficiently large and random networks, the mean degree of a neighbour is given by

$$\sum_k k \times \frac{k p(k)}{\sum_{k'} k' p(k')} = \frac{\langle k^2 \rangle}{\langle k \rangle}. \tag{3.12}$$

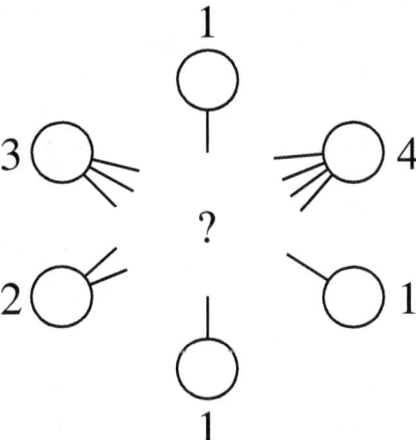

Fig. 3.2 Friendship paradox. A network with $N = 6$ nodes is shown. The numbers represent the nodes' degrees. The expected degree of a neighbour of a randomly selected node is equal to ≈ 2.67, which is larger than the mean degree of the network, $\langle k \rangle = 2$.

3.3   Measures derived from walks and paths

A walk is defined as a succession of adjacent nodes such that one can travel from the start node to the end node by traversing links. A path is a walk where each node is visited only once (with the possible exception that the walk may end at the node where it begins). Walks are used for constructing dynamical processes on networks (e.g., random walks) and measurements such as the Katz centrality, where all possible walks from one node to another are exhaustively counted. Paths are particularly useful when considering the shortest travelling route from one node to another. The situation is pretty much the same for temporal networks introduced in Chapter 4. In the network science literature, which is rooted in statistical physics, authors tend not to distinguish walks and paths. This is particularly the case for the literature on temporal networks. In this chapter, we will however distinguish walks and paths. In the subsequent chapters on temporal networks, we sometimes use the term path to refer to a walk when the path is a commonly used term.

The number of walks of a certain length can be obtained from powers of the adjacency matrix. The adjacency matrix provides the number of walks of length 1 between two nodes. In general, the number of walks of length ℓ is given by the elements of A^ℓ.

To identify paths from a node to another requires more effort. The distance between nodes v_i and v_j, denoted by $d(v_i, v_j)$, is defined as the smallest number of hops in a path necessary to go from v_i to v_j. For undirected networks, the distance defined in this way satisfies the axioms that a distance measure should satisfy: non-negativity (i.e., $d(v_i, v_j) \geq 0$), coincidence (i.e., $d(v_i, v_j) = 0$ if and only if $v_i = v_j$), symmetry (i.e., $d(v_i, v_j) = d(v_j, v_i)$) and triangle inequality (i.e., $d(v_i, v_j) \leq d(v_i, v_\ell) + d(v_\ell, v_j)$). For directed networks, the symmetry is broken because a shortest path from v_i to v_j is not generally the same as that from v_j to v_i.

For both undirected and directed networks, $d(v_i, v_j)$ can be calculated by Dijkstra's algorithm. For a fixed v_i, first initialise the distance from v_i by setting $d(v_i, v_i) = 0$ and $d(v_i, v_j) = \infty (j \neq i)$. Second, set $d(v_i, v_j) = 1$, where v_j is a neighbour of v_i. Third, we declare that v_i has been visited. Fourth, consider each neighbour of v_j except v_i. If a neighbour, v_ℓ, has a tentative distance value from v_i larger than two, then we reset it to $d(v_i, v_\ell) = 2$. When all neighbours of v_j are exhausted, we declare that v_j has been visited. Fifth, we select an unvisited node with the smallest tentative distance value and inspect its all neighbours. We repeat the same procedure to determine the distance from v_i to every other node.

For undirected networks, the average distance for a network is defined by

$$L = \frac{2}{N(N-1)} \sum_{i=1}^{N} \sum_{j=1}^{i-1} d(v_i, v_j). \tag{3.13}$$

In many real networks, L is remarkably small as compared to the number of nodes, N. For example, a Facebook network composed of $N \approx 7.2 \times 10^8$ active users with 6.9×10^{10} friendship links yielded $L \approx 4.7$ [Backstrom *et al.* (2012)]. The diameter is defined by

$$D = \max_{u,v \in V} d(u, v). \tag{3.14}$$

In undirected networks, two nodes are said to be connected if there exists a path between them. Connectedness is an equivalence relation because it is reflexive (i.e., v is connected to itself), symmetric (i.e., if u is connected to v, v is connected to u) and transitive (i.e., if u is connected to v and v is connected to w, u is connected to w). Intuitively, a connected component is an island within which one can travel from any node to any other along a path. There is no path between nodes in different components. Connected components impose limitations on any dynamical process taking place on the network. In epidemic processes, for example, the existence of distinct components implies that certain regions of the network are never infected, independently of the model of epidemic dynamics and its parameters.

In directed networks, symmetry is not satisfied because the existence of a path from u to v does not guarantee the existence of a path from v to u. Therefore, the concept of connectedness is more complex, and the notions of strong and weak connectedness are distinguished. Nodes u and v are said to be strongly connected if there exist a path from u to v and a path from v to u. Two nodes u and v are said to be weakly connected if there exists a path between u and v in a network where the direction of the links is discarded. Both strong and weak connectedness is an equivalence relation and induces strongly and weakly connected components, respectively. For example, a strongly connected component is a maximum set of nodes in which each pair of nodes is strongly connected. Strong connectedness implies weak connectedness but not vice versa.

3.4 Clustering coefficient

Empirical networks are quite often abundant in triangles, i.e., mutually connected three nodes. The amount of triangles in a network is quantified by the clustering coefficient [Watts and Strogatz (1998)]. It is defined through the local clustering coefficient defined by

$$C_i \equiv \frac{\text{number of triangles including the } i\text{th node}}{k_i(k_i - 1)/2}, \tag{3.15}$$

which measures the abundance of triangles in the neighbourhood of the ith node, v_i. The denominator gives the normalisation such that $0 \leq C_i \leq 1$. If any pair of the neighbours of v_i is adjacent to each other to form a triangle, $C_i = 1$. If no pair of neighbours of v_i is adjacent to each other such that the neighbourhood of v_i is star-like, $C_i = 0$. The clustering coefficient, denoted by C, is defined as the average of C_i over the network, i.e.,

$$C \equiv \frac{1}{N} \sum_{i=1}^{N} C_i. \tag{3.16}$$

Note that $0 \leq C \leq 1$.

3.5 Spectral properties

A broad range of dynamical and structural properties of networks is characterised by spectral properties of a matrix describing the network. Depending on the problem at hand, we often use the adjacency matrix (denoted by A), the (combinatorial) Laplacian matrix (denoted by L) or the normalised Laplacian matrix (denoted by $\tilde{L}$). Spectral properties of networks have been studied in detail, and various bounds are available. For detailed discussion, we refer to [Chung (1997); Arenas *et al.* (2008); Cvetković *et al.* (2010); Newman (2010)]. In this section, we present standard methods to estimate eigenvalues and a summary of basic spectral properties of undirected networks.

The Laplacian and normalised Laplacian are defined by

$$L_{ij} = k_i \delta_{ij} - A_{ij}, \tag{3.17}$$

$$\tilde{L}_{ij} = \delta_{ij} - \frac{A_{ij}}{\sqrt{k_i k_j}}. \tag{3.18}$$

The three matrices are connected by the following relationships:

$$L = D - A, \tag{3.19}$$

$$\tilde{L} = D^{-1/2} L D^{-1/2} = I - D^{-1/2} A D^{-1/2}, \tag{3.20}$$

where D is the $N \times N$ diagonal matrix whose (i, i) element is equal to k_i and I is the $N \times N$ identity matrix. Both Laplacian matrices are symmetric and their eigenvectors $\boldsymbol{u}_\ell$ ($1 \leq \ell \leq N$) form an orthonormal basis such that $\langle \boldsymbol{u}_\ell, \boldsymbol{u}_{\ell'} \rangle = \delta_{\ell\ell'}$. Any N-dimensional vector $\boldsymbol{x}$ can be decomposed as

$$\boldsymbol{x} = \sum_{\ell=1}^{N} a_\ell \boldsymbol{u}_\ell, \tag{3.21}$$

where

$$a_\ell = \langle \boldsymbol{x}, \boldsymbol{u}_\ell \rangle. \tag{3.22}$$

For the adjacency matrix, it is customary to order the eigenvectors from the largest λ_1 to the smallest λ_N, whereas the eigenvalues are usually ordered from the smallest to the largest for the Laplacian matrices. The two Laplacian matrices always have a zero eigenvalue. In fact, the corresponding eigenvector for L and $\tilde{L}$ is given by $\boldsymbol{u}_1 = (1 \; \cdots \; 1)^\top$ and $\boldsymbol{u}_1 = (\sqrt{k_1} \; \cdots \; \sqrt{k_N})^\top$, respectively. In undirected networks, the zero eigenvalue, $\lambda_1 = 0$, is an isolated eigenvalue and all the other eigenvalues are positive such that $0 = \lambda_1 < \lambda_2 \leq \cdots \leq \lambda_N$ if the network is connected. In this case, the smallest nonzero eigenvalue of the Laplacian matrix, λ_2, determines the relaxation time of diffusion and synchronisation dynamics induced by L, as in our discussion on Markov chains, and is often called the spectral gap. The corresponding eigenvector, $\boldsymbol{u}_2$, is called the Fiedler vector. In general, the number of connected components is given by the number of zero eigenvalues of L or $\tilde{L}$. Therefore, the network is connected if and only if $\lambda_2 > 0$.

The eigenvectors of the three matrices are the same for regular networks. For regular networks, the eigenvalues of the three matrices are related by

$$\lambda_i(L) = \langle k \rangle - \lambda_i(A), \tag{3.23}$$

$$\lambda_i(\tilde{L}) = 1 - \frac{\lambda_i(A)}{\langle k \rangle}, \tag{3.24}$$

where the argument specifies the matrix. For non-regular networks, these matrices have different spectral properties.

Bounding the eigenvalues in terms of easily accessible quantities such as the mean or maximum degree often helps us to develop theory and facilitate intuitive understanding of networks. Let us first focus on the adjacency matrix. Its eigenvalues are estimated with the use of the Rayleigh coefficient defined for any non-zero vector x by

$$\frac{x^\top A x}{x^\top x} = \sum_{\ell=1}^{N} a_\ell^2 \lambda_\ell. \tag{3.25}$$

We obtain lower and upper bounds by replacing λ_ℓ in the summation either by λ_1 or λ_N, which provides

$$\lambda_N \leq \frac{x^\top A x}{x^\top x} \leq \lambda_1. \tag{3.26}$$

Different choices of x offer us different bounds for these extreme eigenvalues. Substitution of $x = (1 \cdots 1)^\top$ in Eq. (3.26) yields

$$\lambda_N \leq \langle k \rangle \leq \lambda_1. \tag{3.27}$$

The bounds improve by a better choice of x. For instance, let the $i_{\max}$th node have the largest degree in the network. Substitution of $x = (x_1 \cdots x_N)^\top$, where

$$x_i = \begin{cases} \sqrt{k_{\max}} & \text{for } i = i_{\max}, \\ 1 & \text{for neighbours of the } i_{\max}\text{th node}, \\ 0 & \text{otherwise}, \end{cases} \tag{3.28}$$

in Eq. (3.26) yields

$$\sum_{j=1}^{N} A_{ij} x_j \begin{cases} = k_{\max} & \text{for } i = i_{\max}, \\ \geq \sqrt{k_{\max}} & \text{for neighbours of the } i_{\max}\text{th node}, \\ \geq 0 & \text{otherwise}. \end{cases} \tag{3.29}$$

The inequalities are due to the fact that we have only counted the links between the $i_{\max}$th node and its neighbours. The equality in Eq. (3.29) holds true in the case of a star graph. By combining Eqs. (3.28) and (3.29), we obtain

$$\sum_{j=1}^{N} A_{ij}x_j \geq \sqrt{k_{\max}}x_i. \tag{3.30}$$

By multiplying x_i to the left of Eq. (3.30), summing over i and using Eq. (3.26), we obtain

$$\sqrt{k_{\max}} \leq \lambda_1. \tag{3.31}$$

Similar inequalities can be obtained for the two Laplacian matrices. The spectra of the normalised Laplacian, $\tilde{L}$, satisfies

$$0 = \lambda_1 \leq \lambda_2 \leq \cdots \leq \lambda_N \leq 2. \tag{3.32}$$

The largest eigenvalue verifies the equality $\lambda_N = 2$ if and only if the network is bipartite. Useful bounds for the unnormalised Laplacian, L, are:

$$\frac{4}{Nd_{\max}} \leq \lambda_2 \leq \frac{N}{N-1}k_{\min}, \tag{3.33}$$

$$\frac{N}{N-1}k_{\max} \leq \lambda_N \leq 2k_{\max}, \tag{3.34}$$

where $k_{\min}$ and $k_{\max}$ are the minimum and maximum degrees in the network, respectively, and $d_{\max}$ is the diameter of the network.

3.6   Discrete-time random walks on networks

We postpone proper discussion of dynamical processes on networks until we explain them for temporal networks in Chapter 6. Nevertheless, we make an exception here to introduce discrete-time random walks on networks because they are used in various places in this book.

Let us consider a walker diffusing on an undirected network. At each step, the walker located on a node selects one link connected to the node at random and jumps to an adjacent node. This process is equivalent to a Markov chain (Section 2.9) and is described by the $N \times N$ transition matrix T (Eq. (2.104)). The probability that the walker visits the ith node after t steps, $p_i(t)$, is given in by Eq. (2.108), where $\boldsymbol{p}(t) = (p_1(t) \,\cdots\, p_N(t))$.

The solution, Eq. (2.108), involves products of matrices and can be simplified with the use of a graph Fourier transform. The underlying idea is to decompose the signal in an adequate base of vectors, such that the matrix products take the form of algebraic products for amplitudes associated to the different dynamical modes. To work out this idea, let us first note that the transition matrix of the random walk is given by $T_{ij} = A_{ij}/k_i$, representing the probability that the walker transits from the ith node to the jth node. The transition matrix, T, is in general asymmetric, except if the underlying graph is regular. Nonetheless, its spectral properties can be directly derived from those of the symmetric matrix

$$\tilde{A}_{ij} = \frac{A_{ij}}{\sqrt{k_i k_j}}, \tag{3.35}$$

whose properties are essentially equivalent to those of the normalised Laplacian (Eq. (3.18)). By applying the decomposition given by Eq. (2.93) to $\tilde{A}$, we obtain

$$\tilde{A}_{ij} = \sum_{\ell=1}^{N} \lambda_\ell \boldsymbol{u}_\ell \boldsymbol{u}_\ell^\top, \tag{3.36}$$

where λ_ℓ is the ℓth eigenvalue of $\tilde{A}$ and $\boldsymbol{u}_\ell$ is the normalised eigenvector such that $\langle \boldsymbol{u}_\ell, \boldsymbol{u}_{\ell'} \rangle = \delta_{\ell\ell'}$.

Because $T_{ij} = \sqrt{k_j} \tilde{A}_{ij}/\sqrt{k_i}$, i.e., $T = D^{-1/2} \tilde{A} D^{1/2}$, where D is the $N \times N$ diagonal matrix whose (i, i) element is equal to k_i ($1 \le i \le N$), the ℓth left and right eigenvectors of T are given by

$$\boldsymbol{u}_\ell^{\mathrm{L}} = \left((u_\ell)_1 \sqrt{k_1} \ \cdots \ (u_\ell)_N \sqrt{k_N} \right), \tag{3.37}$$

$$\boldsymbol{u}_\ell^{\mathrm{R}} = \left((u_\ell)_1/\sqrt{k_1} \ \cdots \ (u_\ell)_N/\sqrt{k_N} \right)^\top, \tag{3.38}$$

respectively. Equation (3.37) is verified by

$$\left(\boldsymbol{u}_\ell^\top D^{1/2} \right) T = \left(\boldsymbol{u}_\ell^\top D^{1/2} \right) \left(D^{-1/2} \tilde{A} D^{1/2} \right)$$
$$= \left(\boldsymbol{u}_\ell^\top \tilde{A} \right) D^{1/2} = \lambda_\ell \boldsymbol{u}_\ell^\top D^{1/2}, \tag{3.39}$$

which implies $\boldsymbol{u}_\ell^{\mathrm{L}} = \boldsymbol{u}_\ell^\top D^{1/2}$ with eigenvalue λ_ℓ. Equation (3.38) is verified by

$$T \left(D^{-1/2} \boldsymbol{u}_\ell \right) = \left(D^{-1/2} \tilde{A} D^{1/2} \right) \left(D^{-1/2} \boldsymbol{u}_\ell \right)$$
$$= D^{-1/2} \left(\tilde{A} \boldsymbol{u}_\ell \right) = \lambda_\ell D^{-1/2} \boldsymbol{u}_\ell, \tag{3.40}$$

which implies $\boldsymbol{u}_\ell^{\mathrm{R}} = D^{-1/2} \boldsymbol{u}_\ell$ with eigenvalue λ_ℓ.

Using Eq. (2.95), we obtain

$$T^t = \left(D^{-1/2}\tilde{A}D^{1/2}\right)^t = D^{-1/2}\tilde{A}^t D^{1/2}$$

$$= D^{-1/2}\sum_{\ell=1}^{N}\lambda_\ell^t \boldsymbol{u}_\ell \boldsymbol{u}_\ell^\top D^{1/2}$$

$$= \sum_{\ell=1}^{N}\lambda_\ell^t \boldsymbol{u}_\ell^{\mathrm{R}}\boldsymbol{u}_\ell^{\mathrm{L}}. \tag{3.41}$$

Therefore,

$$\boldsymbol{p}(t) = \boldsymbol{p}(0)T^t = \sum_{\ell=1}^{N}\lambda_\ell^t \boldsymbol{u}_\ell^{\mathrm{L}}\langle \boldsymbol{p}(0), \boldsymbol{u}_\ell^{\mathrm{R}}\rangle. \tag{3.42}$$

In Eq. (3.42), $a_\ell(0) \equiv \langle \boldsymbol{p}(0), \boldsymbol{u}_\ell^{\mathrm{R}}\rangle$ is the projection of the initial condition to the ℓth eigenmode. Equation (3.42) indicates that the state of the random walk after t steps is given by a linear combination of the eigenmodes as follows:

$$p_i(t) = \sum_{\ell=1}^{N} a_\ell(t)(u_\ell^{\mathrm{L}})_i, \tag{3.43}$$

where

$$a_\ell(t) = \lambda_\ell^t a_\ell(0). \tag{3.44}$$

As mentioned in Section 2.9, the eigenvalues λ_ℓ of the transition matrix are in the interval $[-1, 1]$. Modes satisfying $|\lambda_\ell| < 1$ relax to zero, and the stationary state is associated with the mode with eigenvalue unity. By definition, the graph Fourier transform maps a vector defined on the nodes, i.e., $\boldsymbol{p}(t)$, to a vector of amplitudes of the eigenvectors $(a_1(t) \cdots a_N(t))$ [Tremblay and Borgnat (2014)]. It is a generalisation of the standard Fourier transform to topologies different from lattices (e.g., one-dimensional chain, two-dimensional square grid), where the eigenvectors of the transition matrix play the role of e^{ikx}. Graph Fourier transforms can also be defined based on other matrices describing the graph, e.g., the adjacency matrix. The choice of matrix and its associated base of eigenvectors are motivated by their adequacy for the problem at hand.

3.7   Centrality

Centrality measures aim to quantify the importance of nodes in a network. The simplest one is the degree (i.e., degree centrality), with which hubs are considered to be important. The degree centrality is effective in various situations but not always. This observation has motivated the introduction of different types of centrality measures. In this section, we explain some of them, which are relatively popular and employed in the context of temporal networks.

3.7.1 Closeness centrality

The closeness centrality and betweenness centrality are popular centrality measures based on the distance between pairs of nodes. The closeness centrality for node v_i is defined by

$$\text{closeness}_i = \frac{N-1}{\sum_{j=1;j\neq i}^{N} d(v_i, v_j)}, \tag{3.45}$$

which is the inverse of the mean distance from node v_i to any other node. The closeness centrality is well-defined only for connected networks.

3.7.2 Betweenness centrality

The betweenness centrality is defined as the fraction of the shortest paths passing through the node in question. This quantity is averaged over all possible pairs of nodes. The betweenness of the ith node is defined by

$$\text{betweenness}_i = \frac{2}{(N-1)(N-2)} \sum_{j=1;j\neq i}^{N} \sum_{\ell=1;\ell\neq i}^{j-1} \frac{\sigma_{j\ell}^i}{\sigma_{j\ell}}, \tag{3.46}$$

where $\sigma_{j\ell}$ is the number of the shortest paths connecting the jth and ℓth nodes, and $\sigma_{j\ell}^i$ is the number of such shortest paths that pass through the ith node. The convention is that we regard the summand on the right-hand side of Eq. (3.46) to be zero when $\sigma_{j\ell}$ is equal to zero (i.e., when the jth and ℓth nodes are in different connected components). The summation excludes the shortest paths that start or end at the ith node because it is obvious that such a path does not go through the ith node. The normalisation factor $2/[(N-1)(N-2)]$ comes from the combinations of j and ℓ, whereas it is often neglected.

3.7.3 Katz centrality

Given an adjacency matrix, A, the number of walks from the ith node to the jth node with ℓ steps is given by the (i, j) element of A^ℓ. Supposing that short walks are more important than long walks in mediating, e.g., communication and infectious diseases, we scale the importance of each walk of length ℓ ($\ell \geq 0$) by a factor of α^ℓ, where $0 < \alpha < 1$. Then, the weighted sum of the number of walks from the ith to the jth nodes of various lengths is given by the (i,j) element of

$$I + \alpha A + \alpha^2 A^2 + \cdots = (I - \alpha A)^{-1}. \tag{3.47}$$

Note that the walks of length zero also contribute to the counting with weight one.

The Katz centrality [Katz (1953)] of the ith node is defined by

$$\mathrm{Katz}_i = \sum_{j=1}^{N} \left[(I - \alpha A)^{-1} \right]_{ij}. \tag{3.48}$$

In other words, the weighted sum of the number of walks starting from the ith node is summed over all destination nodes. If $\alpha = 0$, then $\mathrm{Katz}_i = 1$ for all i. Therefore, we are interested in making α large to diversify the values of Katz_i. In fact, as intuitively understood from Eq. (3.47), $(1 - \alpha A)^{-1}$ diverges for a large α. This occurs when an eigenvalue of $I - \alpha A$ hits zero for the first time as α is increased. Therefore, the Katz centrality is well-defined when α is smaller than the inverse of the largest eigenvalue of A.

3.7.4   *PageRank*

A well-known centrality measure for directed networks is the PageRank, which was first introduced for ranking webpages [Brin and Page (1998); Langville and Meyer (2004, 2006)] and later adopted in a variety of applications. The PageRank is defined as the stationary density of a discrete-time random walk, particularly on directed networks. We studied discrete-time random walks on undirected networks in Section 3.6. In this section, we start by looking at discrete-time random walk on directed networks, which is a representative Markov chain (Section 2.9).

Consider a directed network and a random walker in discrete time. In each step, the walker located at the ith node jumps to one of the out-neighbours selected at random. The transition matrix, i.e., the probability that the walker moves from the ith node to the jth node is given by

$$T_{ij} = \frac{A_{ij}}{k_i^{\mathrm{out}}}. \tag{3.49}$$

Although we focus here on the case of unweighted networks, the following analysis can be easily generalised to the weighted case.

The time evolution of the density of the random walk is driven by Eq. (2.106). The stationary density given by Eq. (2.109) essentially defines the PageRank. The PageRank states that node v_i is important if v_i receives many links, the links entering v_i emanate from important nodes, and a node v_j sending a directed link to v_i has a small out-degree. The last condition says that the total importance of v_j is shared among its out-neighbours. This circular relationship, i.e., a node is important if it is connected to important nodes, leads to an eigenvalue problem.

The naive use of the stationary density $\boldsymbol{p}^* = (p_1^* \cdots p_N^*)$, $p_i^* = \lim_{t\to\infty} p_i(t)$ $(1 \leq i \leq N)$ is in general not appropriate because $\boldsymbol{p}^*$ is not unique when there are multiple absorbing states and the stationary density is equal to zero for transient nodes (Section 2.9). The stationary state uniquely exists if and only if the network is strongly connected, which is rare in empirical directed networks. To circumvent these problems, mathematical tricks have been proposed to make the dynamics ergodic even when the underlying network is not strongly connected. The most popular method consists in allowing walkers to randomly teleport to other nodes. Random walks with teleportation are driven by the rate equation

$$p_i(t+1) = \alpha \sum_{j=1}^{N} p_j(t) T_{ji} + (1-\alpha) u_i, \tag{3.50}$$

where the preference vector $(u_1 \cdots u_N)$, subject to the constraint $\sum_{i=1}^{N} u_i = 1$, determines the probability with which a walker teleports to the ith node when it does. The probability of teleportation is equal to $1 - \alpha$. In the case of web browsing, teleportation is interpreted as a jump to a new webpage without following a hyper-link. In general, if $u_i \neq 0$ $(1 \leq i \leq N)$, any $0 \leq \alpha < 1$ makes the altered random walk given by Eq. (3.50) ergodic such that it converges to a unique stationary state. Use of a small α value makes the convergence to the stationary state faster and numerically stable, but waters down the effect of the network structure encoded in matrix T. A rule of thumb is to set α close to unity, in order to minimise the effect of teleportation, but not too close. A common choice is $\alpha = 0.85$ and $u_i = 1/N$ $(1 \leq i \leq N)$.

The stationary state of Eq. (3.50) is formally given by

$$p_{i;\alpha}^* = (1-\alpha) \sum_{j=1}^{N} u_j \left[(I - \alpha T)^{-1} \right]_{ji}, \tag{3.51}$$

where the dependence of the stationary density on α has been made explicit. This solution can be Taylor expanded in terms of α to yield [Boldi *et al.* (2005); Brinkmeier (2006)]

$$p_{i;\alpha}^* = u_i + \sum_{\ell=1}^{\infty} \alpha^\ell \sum_{j=1}^{N} u_j \left(T_{ji}^\ell - T_{ji}^{\ell-1} \right). \tag{3.52}$$

This expression clearly shows the non-local nature of the PageRank because it is made of walks of all length ℓ. A large α value gives a high credit to long walks. As in the case of the Katz centrality, the stationary density may radically change when α is modified [Langville and Meyer (2004)]. This dependence is clear when rewriting Eq. (3.52) with $u_i = 1/N$ in the following form:

$$p_{i;\alpha}^* = \frac{1}{N} + \sum_{\ell=1}^{\infty} \frac{\alpha^\ell}{N} \sum_{j,j'=1}^{N} \left(\frac{k_{j'}^{\text{in}} - k_j^{\text{out}}}{k_{j'}^{\text{in}}} \right) T_{jj'} T_{j'i}^{\ell-1}. \tag{3.53}$$

The leading contribution for small α makes the PageRank uniform, thereby making all nodes equivalent. Differentiation emerges when α is increased. The contribution of each walk of length ℓ is proportional to $k_{j'}^{\text{in}} - k_j^{\text{out}}$. Note that each term of the summation vanishes when the network is regular, i.e., $k_i^{\text{in}} = k_i^{\text{out}} = M/N$ ($1 \le i \le N$), which yields $p_{i;\alpha}^* = 1/N$ ($1 \le i \le N$) regardless of the α value.

A way to minimise the dependence of the PageRank on α is to carefully choose a preference vector. A choice is $u_i = k_i^{\text{in}}/M$ [Lambiotte and Rosvall (2012)], inspired by the observation that the in-degree of a node is positively correlated with the PageRank in sufficiently random directed networks [Fortunato *et al.* (2008)]. In this case, teleportation is equivalent to selecting a link at random with probability $1/M$ and following it, instead of selecting a node with probability $1/N$. This preference vector results in

$$p_{i;\alpha}^* = \frac{k_i^{\text{in}}}{M} + \sum_{\ell=1}^{\infty} \frac{\alpha^\ell}{M} \sum_{j=1}^{N} \left(k_j^{\text{in}} - k_j^{\text{out}} \right) T_{ji}^\ell, \tag{3.54}$$

which differs from Eq. (3.53) in several ways. At the zeroth order in α, the PageRank for this link teleportation is given by the in-degree of the node, which is itself a popular centrality measure for directed networks. The ℓth order contributions are made of a weighted average of the walks of length ℓ. The contribution of each node in Eq. (3.54), instead of each link in Eq. (3.53), is the difference between its in-degree and out-degree, $k_j^{\text{in}} - k_j^{\text{out}}$. Nodes concentrating the flow of probability, $k_j^{\text{in}} > k_j^{\text{out}}$, have a positive contribution, while nodes diluting the flow, $k_j^{\text{in}} < k_j^{\text{out}}$, have a negative contribution. Equation (3.54) is advantageous in being independent of α when the network is undirected or when the network is Eulerian ($k_i^{\text{in}} = k_i^{\text{out}}$ for each i). In these cases, Eq. (3.54) reduces to $p_i^* = k_i^{\text{in}}/M$.

We have implicitly assumed that each node has at least one outgoing link, i.e., $k_i^{\text{out}} > 0$ ($1 \leq i \leq N$). If this condition is violated, the transition probability given by Eq. (3.49) is ill-defined. Therefore, we usually force a teleportation step with probability unity (not with probability $1 - \alpha$) when a walker arrives at a dangling node, i.e., a node without outgoing links. Mathematically, we set $T_{ji} = u_i$ ($1 \leq i \leq N$) for dangling nodes v_j.

Because the PageRank is the eigenvector corresponding to the largest eigenvalue of a positive matrix T', whose (i, j) element is given by $T'_{ij} = \alpha T_{ij} + (1 - \alpha)u_j$, we can efficiently compute it using the power method (Section 2.7). The power method converges rapidly if the spectral gap of T' is large.

3.8 Models of networks

When analysing structural patterns of empirical networks, it is important to compare their properties with those of appropriate reference points, often produced by models of networks. We distinguish two families of models. The first is random graph models in which links are random variables with certain constraints. The most fundamental model of random graph is the Erdős-Rényi model (Section 3.8.1), and other examples include the configuration model (Section 3.8.2) and the exponential random graph models (Section 5.6). These models provide neither an explanation for the values taken by the parameters nor the reason for certain constraints to emerge in an empirical network. Instead, they have nice mathematical and statistical properties. For this reason, random graphs provide a useful baseline, or null model, for deciding whether patterns observed in empirical data are significant. In practice, if a value of a measurement observed in empirical data is significantly different from the expected value for the random graph model, the model does not represent the process behind the empirical data.

The second class of models is mechanistic models, whose goal is to understand the mechanisms leading to certain structures observed in empirical networks. In general, such models are defined by simple rules on how nodes and links are created or destroyed in the course of time. Examples include the growing network model (Section 3.8.3). Comparison between networks generated by the model and empirical networks allows us to identify potential forces having driven the evolution of the empirical networks.

3.8.1 *Erdős-Rényi random graph*

One of the simplest random graph models is the Erdős-Rényi random graph, introduced by Hungarian mathematicians Erdős and Rényi in 1959 (see [Bollobás (2001)] for detailed exposure). The model, also called the Poisson or binomial random graph, is denoted by $\mathcal{G}(N, q)$ and has two parameters, the number of nodes, N, and the probability q that a link exists between a pair of nodes. The self-loops are excluded. For each pair of nodes, consider a Bernoulli process that determines whether or not they are connected by a link. In fact, $\mathcal{G}(N, q)$ does not represent a single network, but a random ensemble of them in the probabilistic sense. Any network without multiple edges or self-loops is generated by the random graph $\mathcal{G}(N, q)$ as long as $0 < q < 1$. However, the probability that the model generates a target network, or a similar network, may be tiny. When studying the Erdős-Rényi random graph model or other network models defined as a random ensemble of networks (in fact, most models are so), a major aim is to predict the average behaviour of certain network metrics and, if possible, their variance.

In $\mathcal{G}(N, q)$, every link exists independently with the same probability. Therefore, the probability of generating a network with M links in total is given by the binomial distribution as follows:

$$p(M) = \binom{\frac{N(N-1)}{2}}{M} q^M (1 - q)^{\frac{N(N-1)}{2} - M}, \qquad (3.55)$$

where $N(N-1)/2$ is the maximum total number of links in a network. The expected number of links is given by $qN(N - 1)/2$. Similarly, because a node is independently adjacent to any other node with probability q, the degree distribution is given by

$$p(k) = \binom{N-1}{k} q^k (1 - q)^{N-1-k}, \qquad (3.56)$$

which is the binomial distribution.

The Erdős-Rényi model is usually seen as a model for sparse networks, where the total number of links scales linearly with the number of nodes, N. Equivalently, the average degree of the nodes $\langle k \rangle = q(N - 1)$ should not depend on N. Therefore, we usually employ a small value of q, more precisely, $q \propto 1/N$. In the limit of large networks, where $q = \langle k \rangle / (N - 1)$ is sufficiently small to make $\langle k \rangle$ converge to a positive constant, the binomial degree distribution given by Eq. (3.56) is well approximated by the Poisson distribution

$$p(k) = \frac{\langle k \rangle^k}{k!} e^{-\langle k \rangle}. \qquad (3.57)$$

Several properties of the Erdős-Rényi random graph can be derived thanks to the independence of links. Although difficult to derive, the average distance of the Erdős-Rényi random graph is given by

$$L \approx \frac{\log N}{\log \langle k \rangle} \qquad (3.58)$$

for q larger than $O(\log N/N)$ as $N \to \infty$, which ensures that the network is connected [Bollobás (2001)].

The clustering coefficient is given by

$$C = \frac{\binom{N}{3}q^3}{\binom{N}{3}q^2} = q = \frac{\langle k \rangle}{N-1}. \qquad (3.59)$$

When $\langle k \rangle$ does not depend on N, we obtain $C \to 0$ as $N \to \infty$. This calculation can be generalised to the counts of loops or cliques of larger size, leading to a similar observation: the density of such structures decays to zero as $N \to \infty$. This property has important implications in the study of dynamics on random networks, as the network has a locally tree-like structure. If one explores a generated network around a node, the structure is well approximated by a tree, and finding a cross link between two branches of the tree is extremely rare.

The Erdős-Rényi random graph also exhibits a phase transition. Let us consider the size (i.e., number of nodes) of the largest connected component in the network as a function of the mean degree $\langle k \rangle$. When $\langle k \rangle = 0$, the network is trivially composed of N disconnected nodes. In the other extreme of $\langle k \rangle = N - 1$, each node pair is adjacent such that the network is trivially connected. Between the two extremes, the network does not change smoothly in terms of the largest component size. Instead, a giant component, i.e., a component whose size is the largest and proportional to N, suddenly appears as $\langle k \rangle$ increases, marking a phase transition. In general, phase transitions represent qualitative changes in the behaviour of complex systems due to non-linearity. They imply that small changes of a parameter may have drastic consequences on the organisation and dynamics of a system.

The sudden emergence of a giant component is shown as follows. Consider the probability that a randomly chosen node does not belong to the giant component, denoted by u. If a giant component is absent in the network, we obtain $u = 1$. Otherwise, $u < 1$. By definition, if node v_i does not belong to the giant component, it must not be adjacent to any node v_j that belongs to the giant component. Therefore, for an arbitrary node v_j, node v_i is either not adjacent to v_j, which occurs with probability $1 - q$, or adjacent to v_j with the extra condition that v_j does not belong to the giant component, which occurs with probability qu. The probability that v_i does not belong to the giant component via v_j is thus equal to $1 - q + qu$. Because there are $N - 1$ candidate nodes as v_j, we obtain

$$u = (1 - q + qu)^{N-1}. \tag{3.60}$$

By applying

$$\lim_{N \to \infty} \left(1 - \frac{x}{N}\right)^N = e^{-x} \tag{3.61}$$

to Eq. (3.60), we obtain

$$u = e^{-\langle k \rangle (1 - u)}. \tag{3.62}$$

in the limit $N \to \infty$.

The probability S that a node belongs to the giant component is equal to $1 - u$. Substitution of this relation to Eq. (3.62) yields

$$S = 1 - e^{-\langle k \rangle S}. \tag{3.63}$$

The solutions of Eq. (3.63) are given by the intersection of $y = S$ and $y = 1 - e^{-\langle k \rangle S}$ in $0 \leq S \leq 1$. $S = 0$ is always a solution. Another solution exists if and only if $1 - e^{-\langle k \rangle S}$ grows faster than S at $S = 0$. Because $\mathrm{d}(1 - e^{-\langle k \rangle S})/\mathrm{d}S|_{S=0} = \langle k \rangle$, the critical point above which the giant component emerges is given by $\langle k \rangle = 1$. In the sub-critical regime $\langle k \rangle < 1$, no giant component exists, and the network is composed of a multitude of small components. In the super-critical regime $\langle k \rangle > 1$, a giant component made of $\propto N$ nodes emerges. The transition is continuous at the critical point. At the critical point, the size of the connected components obeys a power-law distribution.

An alternative and probably more intuitive way to show the emergence of the giant component is to adopt an dynamical viewpoint to build components by a branching process. We consider an analogue of infectious disease propagating along links and look at a node v_j that has been infected from its neighbour v_i. How many neighbours, other than v_i, can v_j infect in the next round? Because each link is an independent random process, the average number of newly infected nodes is equal to $q(N - 2) \approx \langle k \rangle$, in the limit

$N \to \infty$. When $\langle k \rangle < 1$, the branching process terminates after a finite number of steps, and the components have a finite size. When $\langle k \rangle > 1$, the average number of new nodes grows exponentially and the branching process never ends with a positive probability. In practice, however, it must end because the network is finite.

Instances of the Erdős-Rényi random graph can be generated in different ways. One can perform a Bernoulli test of probability q on all pairs of nodes, which requires $O(N^2)$ operations. Alternatively, one can draw the degrees of N nodes from the Poisson distribution and build a network using the configuration model introduced in the next section, which respects the degree constraint. The second method is substantially faster for large networks, requiring $O(N + M \log M)$ time.

The Erdős-Rényi random graph plays a fundamental role in network science. Its simple rules allow us to understand real-world phenomena, such as a small average distance L. However, it also produces unrealistic patterns, such as locally tree-like structure and the Poisson degree distribution. In fact, a majority of empirical networks has many triangles (i.e., not tree-like) and a long-tailed degree distribution.

3.8.2 *Configuration model*

The configuration model is a generalisation of the Erdős-Rényi random graph to the case of an arbitrary but given degree of each node. It is used to inspect the effect of heterogeneous degree distributions because it does not have more specific features such as high clustering. The model is defined as a random graph in which all possible configurations appear with the same probability under the constraint that node v_i has degree k_i ($1 \leq i \leq N$). The degree sequence $\{k_i\}$ is often generated by a given degree distribution $p(k)$ under the constraint that the sum of the degrees is an even number to satisfy the handshaking lemma. To generate an instance of the configuration model for a given degree sequence, we first create stubs (half-edges) at each node v_i such that its number is equal to k_i. Then, we randomly select pairs of stubs one by one to connect them as a link as long as the tentatively connected pair does not form a multiple edge or self-loop. In fact, we must avoid the case in which the link creation stops in the middle. For example, if there remain three nodes which have 1, 2, and 3 unused stubs, we have to create three more links because there are six stubs remaining. However, we cannot do that without a self-loop or multiple edge.

Consider a large network generated from a configuration model with a given degree sequence. A stub emanating from v_i is connected to v_j with probability $k_j/2M$ because the number of stubs in the network is equal to $2M$. Because v_i owns k_i stubs, the expected number of links between v_i and v_j is given by

$$A_{ij}^* = \frac{k_i k_j}{2M}, \tag{3.64}$$

where A_{ij}^* represents the statistical average of the adjacency matrix.

When $\langle k^2 \rangle$ is finite, the average distance of the configuration model is given by

$$L = 1 + \frac{\log \frac{N}{\langle k \rangle}}{\log \frac{\langle k^2 \rangle - \langle k \rangle}{\langle k \rangle}} \tag{3.65}$$

for large N [Newman *et al.* (2001)]. For the power-law degree distribution $p(k) \propto k^{-\gamma}$, we obtain

$$L \propto \begin{cases} \log \log N, & (2 < \gamma < 3), \\ \log N / \log \log N, & (\gamma = 3), \\ \log N, & (\gamma > 3), \end{cases} \tag{3.66}$$

[Cohen and Havlin (2003)]. Only when $\gamma > 3$, $\langle k^2 \rangle$ is finite in the limit $N \to \infty$ (Section 2.4) such that Eq. (3.66) is consistent with Eq. (3.65). When $\gamma \leq 3$, L is very small such that the network is called ultra-small-world.

The clustering coefficient is given by [Newman (2010)]

$$\begin{aligned} C &= \sum_{k'=1}^{\infty} \sum_{k''=1}^{\infty} \frac{k'p(k')}{\langle k \rangle} \frac{k''p(k'')}{\langle k \rangle} \frac{(k'-1)(k''-1)}{\langle k \rangle N} \\ &= \frac{\left(\langle k^2 \rangle - \langle k \rangle \right)^2}{\langle k \rangle^3 N}. \end{aligned} \tag{3.67}$$

Clustering coefficient C is small unless the degree distribution is highly heterogeneous, in which case $\langle k^2 \rangle \gg \langle k \rangle^2$.

For the annealed adjacency matrix given by Eq. (3.64), the largest eigenvalue is evaluated as

$$\lambda_1 \approx \frac{\langle k^2 \rangle}{\langle k \rangle}. \tag{3.68}$$

Equation (3.68) diverges for networks with power-law degree distributions, so-called scale-free networks, with degree distribution $p(k) \propto k^{-\gamma}$, $\gamma \leq 3$. In this case, the largest eigenvalue diverges as $N \to \infty$.

3.8.3 *Growing network with preferential attachment*

A broad range of networks grow in time in terms of both the number of nodes, N, and the number of links, M. Examples include citations in science, web graph and networks of airports. Network growth is a type of temporal fluctuations of networks. Understanding mechanisms of network growth definitely helps one to understand temporal dynamics of networks. For example, there is a phenomenon called triadic closure, in which if there are links (v_1, v_2) and (v_2, v_3), then a new link (v_1, v_3) is likely to form, yielding a high clustering coefficient [Granovetter (1973); Kossinets and Watts (2006)]. Triadic closure is a mechanism that can be incorporated in growing network models.

In this section, we study a popular growing network model with the preferential attachment mechanism. We explain this model here, not in a later chapter on temporal network models (Chapter 5), because it plays a pivotal role in the entire network science. The model was proposed by Barabási and Albert [Barabási and Albert (1999)], which we call the BA model, while the model had been known for longer time [de Solla Price (1976); Szymański (1987); Mahmoud *et al.* (1993)]. The model is an instance of a family of multiplicative stochastic models, starting around a century ago with the Pólya urn model and the Yule process. Historically, the mechanism of preferential attachment was also identified qualitatively by the sociologist Robert Merton, who called it the Matthew effect, after a passage in Biblical Gospel of Matthew. The Yule process was studied by the economist Herbert Simon, interested in the distribution of wealth, who showed that it produces power-law distributions. This work inspired the Price's network model [de Solla Price (1976)].

The BA model produces a network according to the following steps:

(1) Prepare m_0 nodes each of whose degree is at least one. A typical choice is the complete graph (i.e., a link exists between every pair of nodes) on m_0 nodes. Set a clock to $t = 0$.
(2) Add a node with $m(\leq m_0)$ half-edges to the existing network. Suppose that the existing network has N' nodes with degrees k_i $(1 \leq i \leq N')$. The probability that each half-edge connects to v_i is specified by

$$\Pi(k_i) = \frac{k_i}{\sum_{j=1}^{N'} k_j} \quad (1 \leq i \leq N'). \tag{3.69}$$

Equation (3.69) indicates that a node receives a new link with the probability proportional to its degree, hence the name of preferential

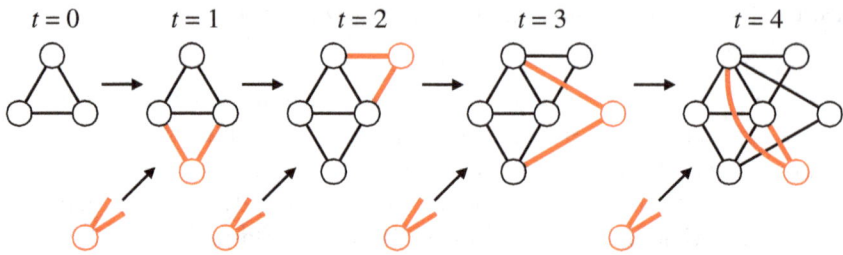

Fig. 3.3 First several stages of the BA model. The bold lines represent new links. We set $m_0 = 3$ and $m = 2$.

attachment. Equation (3.69) is applied under the constraint that we avoid multiple edges, although this constraint is non-essential. When $m \geq 2$, we have to decide on whether or not to update the relevant k_i values used in Eq. (3.69) when one of the m links has been added. However, this decision is again immaterial.

(3) Add nodes one by one until we have N nodes according to step (2). Some of the first stages are schematically shown in Fig. 3.3.

The BA model produces networks with a power-law degree distribution $p(k) \propto k^{-3}$ (Appendix C). In early stages, nodes have similar values of k, which are equal to or slightly larger than m. However, once $\{k_i\}$ becomes somewhat heterogeneous, the heterogeneity will self-reinforce owing to the preferential attachment mechanism.

Several properties of the model have been derived analytically. By allowing self-loops and multiple edges to facilitate mathematical analysis, one obtains [Bollobás and Riordan (2004)]

$$L \propto \begin{cases} \log N, & (m = 1), \\ \log N / \log \log N, & (m \geq 2). \end{cases} \quad (3.70)$$

The network has a small average distance already with $m = 1$, and L is even smaller for $m \geq 2$. The clustering coefficient is given by [Klemm and Eguíluz (2002); Bornholdt and Schuster (2003); Barrat and Pastor-Satorras (2005)]

$$C \approx \frac{m-1}{8} \frac{(\log N)^2}{N}. \quad (3.71)$$

Equation (3.71) implies that, although the BA model has more triangles than the Erdős-Rényi random graph (i.e., $C \propto 1/N$), the BA model still lacks the clustering property because $\lim_{N \to \infty} C = 0$. Various extensions of the BA model realise a non-vanishing C value as $N \to \infty$.

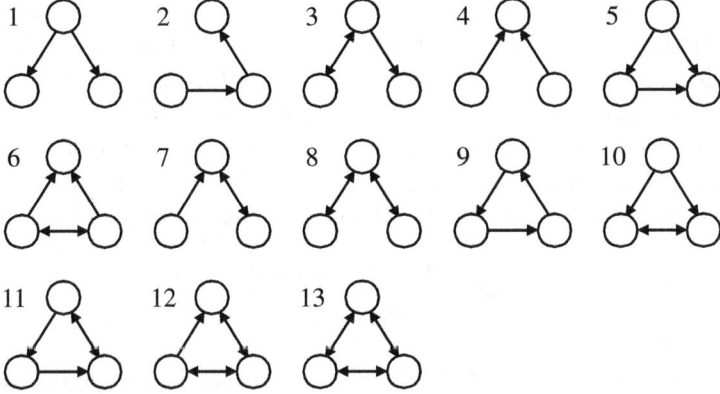

Fig. 3.4   Possible three-node motifs in static directed networks.

3.9   Network motifs

In a majority of empirical networks, triangles are abundant, which is the motivation of measuring the clustering coefficient. We can use the same argument to search for frequent small subgraphs as signatures of a given network. This is motif analysis, a computational method to enumerate small subgraphs (especially, three-node subgraphs) embedded in networks and assess whether a subgraph is significantly frequent [Milo *et al.* (2002, 2004)]. Significantly frequent subgraphs are called network motifs.

If we confine ourselves to weakly connected three-node subgraphs in directed networks, there are 13 candidate network motifs, as shown in Fig. 3.4. We should not determine relative frequency of these subgraphs simply by comparing the number of their appearance. For example, in sparse random networks, a subgraph containing fewer links, such as subgraph 1 in Fig. 3.4, would be more frequently found than a subgraph containing many links, such as subgraph 13. However, the difference in the frequency of subgraph 1 and that of subgraph 13 should be ascribed to sparsity of the network in this case, not to a particular tendency that this network prefers subgraph 1.

Therefore, we measure the frequency of each subgraph relative to that of a random network null model. While different null models do the job, the most frequently used null model is the directed variant of the configuration model, which is a random graph with the in- and out-degree of each node in a given network conserved. The implicit assumption then is that we are interested in overrepresented subgraphs that cannot be explained by heterogeneity in the degree distribution (more precisely, the in- and out-degrees of each node in a given network). For any of the 13 three-node subgraphs, m, we denote by $C(m)$ the frequency of subgraph m in the given network and by $\tilde{C}(m)$ the frequency in a network generated from the configuration model. Because the configuration model is a random network model, the value of $\tilde{C}(m)$ is generally different every time we generate a network from the same configuration model. We define the Z score by

$$Z = \frac{C(m) - \langle \tilde{C}(m) \rangle}{\text{std}\left[\tilde{C}(m)\right]}, \tag{3.72}$$

where $\langle \tilde{C}(m) \rangle$ is the mean of $\tilde{C}(m)$ over the instances of networks generated by the configuration model, and $\text{std}\left[\tilde{C}(m)\right]$ is the standard deviation of $\tilde{C}(m)$. We calculate $\langle \tilde{C}(m) \rangle$ and $\text{std}\left[\tilde{C}(m)\right]$ on the basis of sufficiently many instances to lessen fluctuations in the estimates.

The Z score represents the normalised frequency of subgraph m relative to the configuration model. Calculating the Z score is essentially equivalent to calculating the p value in a statistical test. If the Z score is sufficiently large or small, subgraph m is overrepresented or underrepresented, respectively. Significantly overrepresented subgraphs are network motifs. Although the Z score is famous in the context of network motifs, significance of any quantity measured for a given network should be tested with the Z score whenever possible. Otherwise, we can be easily fooled by intuitively (but not necessarily statistically) large/small values of a measurement.

Network motifs can also be examined with larger subgraphs [Milo *et al.* (2002)] and undirected networks [Milo *et al.* (2004)] although there are exploding numbers of subgraphs to be searched and enumeration of each subgraph is computationally difficult for large subgraphs. For historical reasons, network motifs seem to be a main analysis tool for directed networks. Software mfinder is freely available for finding network motifs [mfinder (2016)].

3.10   Community detection

Many networks exhibit community structure. Community structure implies that the network is composed of groups of nodes such that the nodes are densely connected within the same group and relatively sparsely connected

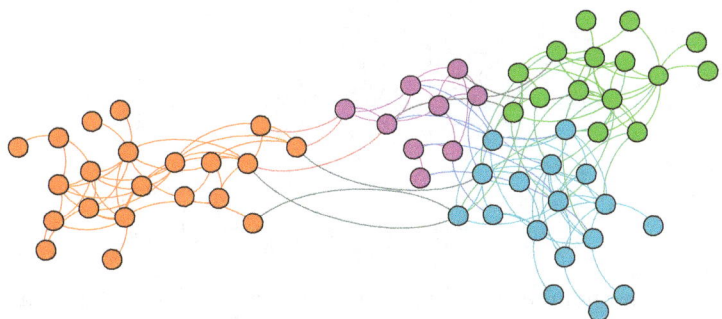

Fig. 3.5 A social network of bottlenose dolphins. Four communities detected by the Louvain algorithm implemented on gephi (http://www.gephi.org) are shown by different colours. Data from [Lesseau *et al.* (2003)].

across different groups. According to a community detection algorithm, the social network of bottlenose dolphins shown in Fig. 3.5 has four communities indicated by different colours. There are many algorithms aiming to detect community structure in a given network in the absence of predefined labelling of nodes [Fortunato (2010)]. In this section, we review some representative approaches to community detection in static networks.

3.10.1 *Modularity*

Modularity, denoted by Q, is a quantity introduced to measure the goodness of the partitioning of a network into communities. This quantity is often used as an objective function to be optimised in order to uncover the best partition of a network. Let us consider a set of nodes, denoted by CM. The underlying idea of modularity is to compare the number of links connecting nodes within CM with the expected number of links in an appropriate null model. Under the configuration model, the probability that nodes v_i and v_j are adjacent is given by $k_i k_j / 2M$ (Eq. (3.64)). Therefore, we quantify the contribution of CM to Q as

$$\sum_{\substack{i,j=1; \\ v_i, v_j \in \text{CM}}}^{N} \left(A_{ij} - \frac{k_i k_j}{2M} \right). \qquad (3.73)$$

Let us now consider a partition of the network into N_{CM} communities. The cth community ($c = 1, 2, \ldots, N_{\text{CM}}$) is denoted by CM_c. Modularity is simply defined as a properly normalised sum of Eq. (3.73) over all communities, i.e.,

$$Q = \frac{1}{2M} \sum_{c=1}^{N_{\text{CM}}} \left[\sum_{\substack{i,j=1; \\ v_i,v_j \in \text{CM}_c}}^{N} \left(A_{ij} - \frac{k_i k_j}{2M} \right) \right]$$

$$= \sum_{c=1}^{N_{\text{CM}}} \left[\frac{M_c}{M} - \left(\frac{\sum_{i=1;v_i \in \text{CM}_c}^{N} k_i}{2M} \right)^2 \right], \qquad (3.74)$$

where M_c is the number of links connecting two nodes within community CM_c. According to Eq. (3.74), calculation of Q only requires the number of intra-community links and the sum of the degree of nodes within each community. We can also rewrite Q as

$$Q = \frac{1}{2M} \sum_{i,j=1}^{N} \left(A_{ij} - \frac{k_i k_j}{2M} \right) \delta(g_i, g_j), \qquad (3.75)$$

where g_i is the community that the ith node belongs to, and $\delta(g_i, g_j) = 1$ if $g_i = g_j$ and $\delta(g_i, g_j) = 0$ otherwise.

The main advantage of modularity over other quality functions for node partitioning is that it allows us to compare partitions made of different numbers of communities. Maximising Q is therefore expected to uncover the best partition into the best number of communities. However, the problem is not that simple, as we will discuss later. Modularity ranges in $[-0.5, 1]$ [Brandes *et al.* (2008)]. The trivial partition into one large community always yields

$$Q = \frac{1}{2M} \sum_{i,j=1}^{N} \left(A_{ij} - \frac{k_i k_j}{2M} \right)$$

$$= \frac{1}{2M} \left(2M - \frac{\sum_{i=1}^{N} k_i \sum_{j=1}^{N} k_j}{2M} \right)$$

$$= \frac{1}{2M} \left(2M - \frac{2M \times 2M}{2M} \right) = 0, \qquad (3.76)$$

where we used the handshaking lemma (Eq. (3.7)).

Because maximisation of Q is an NP-complete problem [Brandes *et al.* (2008)], various algorithms have been developed for calculating approximate maxima of Q with reasonable computational costs. An example is the following greedy algorithm [Newman (2004)].

(1) Partition the given network into $N_{\text{CM}} = N$ communities. Each node constitutes a single community.

(2) Select the pair of communities whose merger maximises Q, i.e., increases it most or decreases it the least as compared to the previous step, and merge them. N_{CM} decreases by one.

(3) Repeat step (2) to merge a pair of communities one by one. In each merger, the merger that maximises Q is adopted.

(4) The algorithm therefore produces N partitions. Among the partitioning results from $N_{CM} = N$ through $N_{CM} = 1$, the final output is the partitioning that has realised the maximum value of Q. The number of communities, N_{CM}, is automatically determined.

We can easily calculate the amount of change in Q upon a merger. Without loss of generality, let us merge communities 1 and 2. The merger only affects the terms with $c = 1$ and $c = 2$ within the summation in Eq. (3.74). Upon the merger, the first term on the right-hand side of Eq. (3.74) increases by

$$\frac{\text{(number of links connecting CM}_1 \text{ and CM}_2 \text{ before the merger)}}{M} \quad (3.77)$$

because these links turn into intra-community links after the merger. The second term changes by

$$\left(\sum_{i=1; v_i \in \text{CM}_1 \text{ or } \text{CM}_2}^{N} \frac{k_i}{2M} \right)^2 - \sum_{c=1}^{2} \left(\sum_{i=1; v_i \in \text{CM}_c}^{N} \frac{k_i}{2M} \right)^2$$

$$= \frac{\left(\sum_{i=1; v_i \in \text{CM}_1}^{N} k_i \right) \times \left(\sum_{i=1; v_i \in \text{CM}_2}^{N} k_i \right)}{2M^2}. \quad (3.78)$$

The total change in Q is given by Eq. (3.77) subtracted by Eq. (3.78), which can be calculated only by counting the number of inter-community links and degree sums. Several variants of this greedy algorithm have been developed, in order to overcome some of its limitations and improve it in terms of the speed and precision. In particular, the so-called Louvain algorithm is widely used [Blondel *et al.* (2008)]. The community structure shown in Fig. 3.5 has been detected by the Louvain algorithm.

Methods based on modularity maximisation suffer from several drawbacks. First, by construction, they are not capable of uncovering overlapping communities often observed in empirical networks (Section 3.10.4). Second, Q exhibits a resolution limit, because using Q it is impossible to detect dense clusters of nodes that are smaller than a certain scale [Fortunato and Barthélemy (2007)]. The resolution limit originates from the dependency of the null model on $2M$. The dependency decreases when the number of links, M, is increased. Then, modularity maximisation tends to favour larger communities. In the limit $M \to \infty$, the null model is neglected and modularity optimisation simply uncovers the connected components. Modularity-based methods implicitly favour communities having a certain size, depending on the size of the entire network, not only on its internal structure. Third, the modularity landscape is usually extremely rugged and degenerate such that there exists an exponential number of alternative, high-scoring partitions [Fortunato (2010); Good *et al.* (2010)]. Finally, although modularity allows us to compare partitions of the same network, it is by no means intended to compare modularity values of different networks. Therefore, Q should not be used as a measure of the modularity of a network. For instance, the modularity of the best partition of a random network tends to $Q = 1$ when the network is sufficiently large, whereas this network is by no means modular [Guimerà *et al.* (2004)].

3.10.2 *Markov stability*

A deeper understanding of modularity, as well as rectification of some of its limitations, is gained from a random walk perspective. Modularity is related to Markov stability, a quality function representing the tendency for a random walker to remain within a community for a long time. More precisely, the Markov stability of a partition is defined as the probability that the walker is in the same community initially and after time t in the stationary state [Lambiotte *et al.* (2009)]. Markov stability can be defined for different dynamical processes, each one giving rise to a different quality function and, in principle, to a different optimal partition of a network.

We focus on a continuous-time random walk, where the walker travels from a node to another as in the discrete-time counterpart (Section 3.6) and the transition rate for the walker is set to unity. A general description of continuous-time random walks will be given in Section 6.3. Denote by p_i the probability that the random walker visits node v_i, omitting the time variable t. The master equation for the random walk process is given by

$$\frac{\mathrm{d}\boldsymbol{p}}{\mathrm{d}t} = -\boldsymbol{p}L, \qquad (3.79)$$

where $\boldsymbol{p} = (p_1, \ldots, p_N)$ and L is the so-called random walk normalised Laplacian matrix whose elements are given by

$$L_{ij} = \delta_{ij} - \frac{A_{ij}}{k_i}. \qquad (3.80)$$

It should be noted that the probability with which a walker at the ith node moves to the jth node is given by $T_{ij} = A_{ij}/k_i$. The steady state is given by the left eigenvector of L corresponding to the zero eigenvalue. Assuming that the network is undirected and connected, direct substitution verifies that the stationary density is given by

$$p_i^* = \frac{k_i}{\sum_{\ell=1}^{N} k_\ell} = \frac{k_i}{2M} \quad (1 \le i \le M), \tag{3.81}$$

where the last equality holds true owing to the handshaking lemma (Eq. (3.7)).

To define Markov stability, denoted by $R(t)$, consider a pair of nodes v_i and v_j belonging to the same community. Equation (3.79) implies that, at stationarity, the joint probability that the walker visits v_i at time 0 and v_j at time t is given by $p_i^*(e^{-tL})_{ij}$. As in the case of modularity, this quantity has to be compared with an appropriate null model. For Markov stability, the null model is given by the joint probability of finding a first walker on v_i at time 0 and a second independent walker on v_j at time t, i.e., $p_i^* p_j^*$. The null model is also regarded as the probability $p_i^*(e^{-tL})_{ij}$ as $t \to \infty$, because the position of a walker at large times is independent of its initial position. By comparing the actual and independent cases, we define

$$R(t) = \sum_{i,j=1}^{N} \left[\left(p_i^* e^{-tL} \right)_{ij} - p_i^* p_j^* \right] \delta(g_i, g_j). \tag{3.82}$$

Markov stability differs from modularity in several ways. First, it involves the exponential of the Laplacian, $e^{-tL} = I - tL + t^2 L^2/2 + \cdots$ and thus combines paths of all lengths between two nodes. Second, Markov stability naturally incorporates a resolution parameter t. A larger value of t gives a more weight to longer paths in the exponential, corresponding to an exploration of the network at a larger time scale. The time thus acts as a resolution parameter enabling us to zoom in and out to uncover the multiscale structure of the network. The optimal partition based on Markov stability is made of less, larger communities when t is increased. Third, Markov stability is a method derived from flows of probability. It is not a combinatorial method like modularity, in which finite elements (links) are counted. For this reason, Markov stability can detect the impact that certain types of network structure may have on dynamical processes. Finally, Markov stability has desirable mathematical properties including connections to spectral graph theory. It is exactly optimised by the bipartition given by the signs of the Fiedler vector, i.e., the second dominant eigenvector of the Laplacian, when t is sufficiently large.

In practice, estimating the exponential of the Laplacian for a large network can be computationally expensive. Therefore, it is often preferable to study a linearised version of stability, defined by its Taylor expansion for small t. Using $(e^{-tL})_{ij} \approx \delta_{ij} - tL_{ij}$, Eq. (3.80), $p_i^* = k_i/2M$ and $p_j^* = k_j/2M$, we reduce Eq. (3.82) to

$$R(t) = \frac{1}{2M} \sum_{i,j=1}^{N} \left[tA_{ij} + (1-t)\delta_{ij}k_i + \frac{k_ik_j}{2M} \right] \delta(g_i, g_j). \qquad (3.83)$$

Because $\sum_{i,j=1}^{N}(1-t)\delta_{ij}k_i\delta(g_i, g_j) = \sum_{i=1}^{N} k_i$ is independent of the partitioning of the network, maximisation of $R(t)$ is equivalent to maximisation of

$$Q(\gamma) = \frac{1}{2M} \sum_{i,j=1}^{N} \left(A_{ij} - \gamma\frac{k_ik_j}{2M} \right) \delta(g_i, g_j), \qquad (3.84)$$

where we introduced the structural resolution parameter $\gamma = 1/t$, more commonly used than t. We remove the condition that t is small.

With $\gamma = 1$, Eq. (3.84) coincides with Eq. (3.75), such that modularity can be seen as a simple, approximate instance of Markov stability. A large value of γ emphasises the penalty in classifying nodes in the same group such that it yields a large number of communities. A small value of γ yields a small number of communities. Resolution parameter γ allows us to tune the characteristic size of the communities in the optimal partition, i.e., not using the characteristic size imposed by modularity maximisation. However, it also raises the important problem of finding a γ value consistent with natural scales of the network [Delvenne *et al.* (2013)].

3.10.3   *Infomap*

In this section, we sketch an alternative community detection method based on random walks, called the Infomap [Rosvall and Bergstrom (2008)]. The method, originally proposed for non-overlapping community structure, works for directed and weighted networks.

Imagine a random walk on a given network. If the network has community structure, the random walker would wander within a community for a long time before crossing a bridge to a different community. A straightforward way to describe the trajectory of the random walk is to write down the visited nodes in an ordered list, e.g., $v_1, v_4, v_1, v_7, v_3, \ldots$. The amount of information required to express the trajectory is estimated as follows. We code each node into a finite binary sequence, i.e., a code word, and concatenate the code words. For example, if v_1, v_3, v_4, and v_7 are coded into 000, 010, 011 and 110. The aforementioned trajectory is coded into $000011000110010\cdots$. For unique decoding, the code has to be prefix-free. In other words, a code word must not be a prefix (i.e., initial segment) of another code word. For example, if v_1 and v_2 are coded into 000 and 0001, respectively, the code is not prefix-free because 000 is an initial segment of 0001.

The Huffman code is a prefix-free code that encodes symbols separately and generally yields short binary sequences to represent trajectories of the random walk. It assigns a short code word to a frequently visited node and vice versa. The mean code word length per step of the random walk is given by $\sum_{i=1}^{N} p_i^* L(i)$, where p_i^* is the stationary density of the random walk at node v_i and $L(i)$ is the length of the code word for node v_i.

When the symbols (v_i in our case) appear independently, the Huffman code often yields a code length that is close to the theoretical lower bound obtained by the Shannon entropy, which is

$$H = -\sum_{i=1}^{N} p_i^* \log p_i^* \qquad (3.85)$$

per step. However, the sequence of nodes is correlated in time because it is produced by the random walk. Then, an alternative coding scheme may lessen the mean code length. In particular, we can design a two-layered variant of the Huffman code to exploit the community structure of the network. Because there are less nodes in a community CM_i as compared to the entire network, we can express a trajectory within CM_i with a shorter, different Huffman code, which is local to CM_i. Based on this observation, we rebuild the Huffman code as follows.

(1) When the walker enters community CM_i, a code word to represent this entry event is issued.

(2) The walker wanders within CM_i. The trajectory of the walker during this period is encoded by concatenating the code words corresponding to the sequence of the visited nodes. The sequence of these code words is simply placed after the code word produced in the previous step (i.e., entry to CM_i). It should be noted that the intra-community code words make sense only within CM_i. A different community $CM_{i'}$ ($i' \neq i$) may use the same code word as the one used within CM_i to represent a different node in $CM_{i'}$.

(3) The walker exits CM_i. This event is represented by a special code word, which is concatenated after the sequence of code words produced so far.

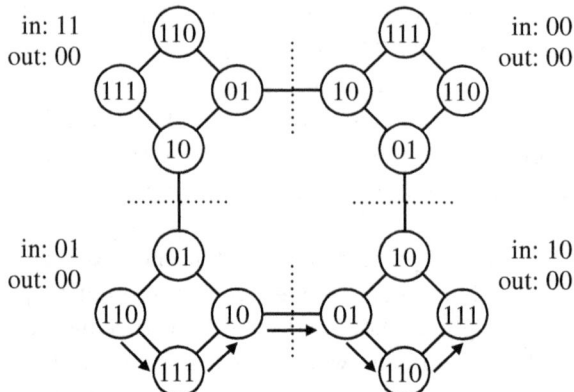

Fig. 3.6 Optimal partitioning according to Infomap and the resulting code words. This example is based on a demo applet available at Martin Rosvall's website http://www.mapequation.org/apps/MapDemo.html.

(4) The exit from CM_i implies that the walker immediately enters a different community, CM_j. Therefore, a code word to notify that the walker has entered CM_j is issued. Then, the code words local to CM_j are used until the walker exits CM_j. We repeat this procedure.

Consider the example shown in Fig. 3.6. The optimal partitioning of this network obtained by Infomap is into four communities whose boundaries are shown by the dotted lines. The binary code word assigned to each node represents the local code within the corresponding community. Note that the code word uniquely determines a node within a community, but not across communities; the same code word is used in different communities. When the random walker enters or exits a community, the corresponding 'in' or 'out' code word is used, respectively. Therefore, the trajectory shown by the arrows in the figure is encoded into 0111011110001001110111. The first 01 indicates that the walk starts in the left-bottom community, and the 110 that follows indicates that the walk starts at the 110 node in this community. 0010 in the middle indicates that the walk exits this community (by the code word 00) and immediately enters the community to the right (by the code word 10).

In contrast to the original Huffman code, we have to invest $2N_{CM}$ code words to mark the entry to and exit from a community. However, we can save the code length when the walker wanders in a community, which occupies a majority of steps. Overall, the mean code length is expected to be

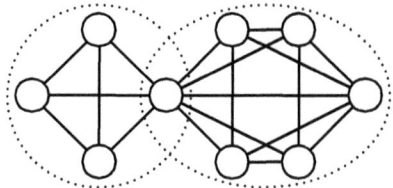

Fig. 3.7 Schematic of overlapping communities. Two communities are shown by dotted circles. One node belongs to both communities.

smaller with the two-layer code in the presence of community structure. In order to detect communities in practice, there is no need for devising the optimal code of a given partition. Infomap instead proceeds by optimising a quality function, called the map equation, which generalises Eq. (3.85). The resulting quality function provides a theoretical limit of how concisely we can specify a walk in the network using a given partition. The optimisation is then performed by a greedy algorithm similar to the one used for maximising modularity (Section 3.10.1), with additional fine- and coarse-graining steps carried out for improving the partitioning.

3.10.4 *Overlapping communities*

So far, we have focused on methods uncovering non-overlapping communities, where nodes are imposed to belong to a single community and a community is defined to be a dense group of nodes. In these methods, we partition the set of nodes into some disjoint groups. However, the field of community detection and more generally of detecting mesoscale structures in networks is much broader. In this short section, we list some approaches.

The concept of overlapping communities is such an extension especially relevant to social networks, in which individuals tend to belong to different social groups. In an example shown in Fig. 3.7, we observe two communities whose boundaries are shown by dotted circles, and a node belongs to both communities. Starting with the k-clique algorithm [Palla *et al.* (2005)], there are now several algorithms to detect overlapping community structure in a network, including methods based on partitioning of links instead of nodes [Evans and Lambiotte (2009); Ahn *et al.* (2010)], optimisation of a quality function for a cover of nodes [Friggeri *et al.* (2011)] and local rather than global optimisation [Lancichinetti *et al.* (2011)].

Chapter 4

Analysis of temporal networks

In this chapter, we start by defining temporal networks, in particular with two representations of them. Then, we introduce various measurements, algorithms and structural properties for temporal networks. A majority of them has counterparts for static networks, which have been explained in Chapter 3. Others do not have static counterparts because consideration of the time induces concepts and measurements that are not relevant to static networks, such as the timing of link usage.

4.1 Definition

Static networks are represented in two main ways: adjacency matrix and link list. Each representation has its own merits and demerits (Section 3.1). The situation is similar for temporal networks. Among various complementary ways to represent temporal networks [Casteigts *et al.* (2011); Holme (2015)], we focus on two representations, which we call the event-based and snapshot representations.

Let us consider a temporal network on N nodes defined over a time span $[0, t_{\max}]$, where $t_{\max}$ is called the lifetime or observation period. Akin to the link list of static networks, the event-based representation of temporal networks consists in viewing the temporal network as a collection of time-stamped links, which we call events. One of the earliest articles on temporal networks also employed the event-based representation [Kempe *et al.* (2000)]. With this representation, a temporal network is given by a time-ordered list of events

$$\{(u_i, v_i, t_i, \Delta t_i); i = 1, 2, \ldots\}, \tag{4.1}$$

where u_i and v_i are the node pair where the ith event occurs, and t_i

and Δt_i are the time of the ith event and its duration, respectively. If a temporal network is directed (e.g., one person sending message to another), we interpret that there is a directed event from u_i to v_i. We let $0 \leq t_1 \leq t_2 \leq \cdots \leq t_n \leq t_{\max}$, where n is the number of events. Within a temporal network, the same node pair (u, v) can appear in multiple events in different times, possibly with different durations. In real data, t_i is often discretised with a resolution of, for example, 20 seconds, due to a finite resolution in observation (e.g., [Cattuto *et al.* (2010)]).

The duration of an event can be quite large [Cattuto *et al.* (2010)]. However, in many situations, Δt_i is very small as compared to the inter-event time, in which case we can ignore the duration of each event. Doing so makes numerical simulations of dynamical processes on temporal networks, as well as building theories, much feasible. In this case, we simplify the event-based representation to

$$\{(u_i, v_i, t_i); i = 1, 2, \ldots\}. \tag{4.2}$$

An example of the event-based representation without the duration of events is shown in Fig. 4.1(a). The network has $N = 4$ nodes, $t_{\max} = 12$ and 14 events scattered on five links, i.e., (v_1, v_3), (v_1, v_4), (v_2, v_3), (v_2, v_4) and (v_3, v_4). For simplicity, in the rest of the book we do not consider the duration of events when we use the event-based representation.

A temporal network can be alternatively represented by a discrete-time sequence of networks

$$\mathcal{G} = \{G(1), G(2), \ldots, G(t_{\max})\}, \tag{4.3}$$

where $t_{\max}$ is now the number of networks. An equivalent, more numeric representation is a sequence of adjacency matrices given by

$$\mathcal{A} = \{A(1), A(2), \ldots, A(t_{\max})\}, \tag{4.4}$$

where $(A(t))_{ij}$ indicates the presence of an event between nodes v_i and v_j at time t ($1 \leq t \leq t_{\max}$). The time in this representation is discretised, whereas the event-based representation allows both continuous and discrete time. We call this representation the snapshot representation of the temporal network, where each $G(t)$, or the corresponding adjacency matrix $A(t)$ when no confusion arises, is called the snapshot. As in the adjacency matrix representation for static networks, the snapshot representation is useful for developing matrix-based theory for temporal networks. We can also view $\mathcal{A}$ as a three-way adjacency tensor indexed by i, j and t, which invites the use of tensor analysis (Section 4.12). This representation emphasises its

(a) (b)

(1,3,1)
(1,4,1)
(2,3,2)
(1,4,4)
(3,4,5)
(2,3,6)
(2,4,6)
(3,4,6)
(1,4,7)
(3,4,9)
(2,4,10)
(3,4,10)
(2,4,11)
(3,4,12)

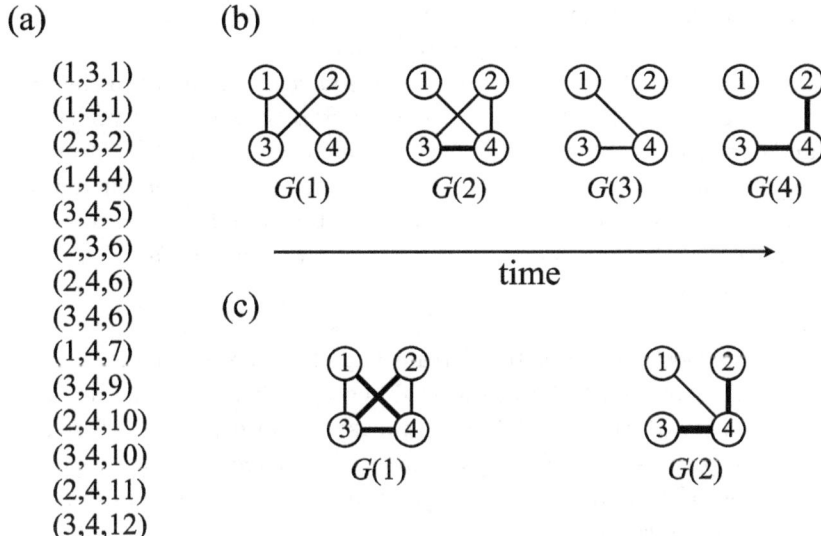

(c)

Fig. 4.1 Two representations of a temporal network. The network has $N = 4$ nodes and the observation period is equal to $t_{\max} = 12$. (a) Event-based representation. (b) Snapshot representation with $T_{\mathrm{w}} = 3$. It should be noted that, by definition, $t_{\max}$ in the snapshot representation is equal to the number of snapshots, i.e., $t_{\max} = 4$. (c) Snapshot representation with $T_{\mathrm{w}} = 6$, for which there are $t_{\max} = 2$ snapshots. The width of lines in (b) and (c) is proportional to the link weight.

macroscopic properties at each time and views the temporal network as a structure evolving in time.

If a temporal network is given in the form of the event-based representation and the time is discretised, it can be transformed to the snapshot representation without loss of information. This is often the case because temporal networks are often observed periodically (e.g., every 20 seconds) for technical reasons. Otherwise, we must specify time windows to coarse-grain an event-based representation of the temporal network into the snapshot representation, and thus lose some temporal information in the original data. If we use a time window of length T_{w} uniformly for all snapshots, the events occurring in $0 \leq t < T_{\mathrm{w}}$ enter snapshot $G(1)$, those occurring in $T_{\mathrm{w}} \leq t < 2T_{\mathrm{w}}$ enter $G(2)$ and so forth. The snapshot representation obtained by coarse-graining the event-based representation shown in Fig. 4.1(a) with $T_{\mathrm{w}} = 3$ is shown in Fig. 4.1(b). We cannot recover Fig. 4.1(a) from Fig. 4.1(b). A larger value of T_{w} causes more loss of information (Fig. 4.1(c)). In the extreme case of $T_{\mathrm{w}} = t_{\max}$, there is just a single snapshot, which is the adjacency matrix of the static network. In this case, all the temporal information is lost.

In static networks, unweighted networks have originally been the major object to be studied until seminal papers emphasised the importance of weighted networks [Yook *et al.* (2001); Barrat *et al.* (2004)]. Even now, unweighted networks seem to be mainly studied. However, in temporal networks, it is a norm rather than exception that a link is used more than once during the time course. Therefore, when we compare a given temporal network with a static counterpart which disregards the time, the static network that we consider is very often weighted. We refer to the static weighted network obtained by disregarding the time of a given temporal network, or equivalently by coarse-graining it with $T_\mathrm{w} = t_\mathrm{max}$, as the aggregate network. Static weighted networks are also popularly used in a snapshot representation of temporal network obtained by coarse graining. For example, $G(2)$ and $G(4)$ in Fig. 4.1(b) and $G(1)$ and $G(2)$ in Fig. 4.1(c) are weighted networks.

Most of the papers use either the event-based or snapshot representation of the temporal network, often without declaration. In the following, we will be explicit about which representation is used as much as possible.

4.2 Temporal walks and paths

4.2.1 *Definition*

The temporal walk is the temporal extension of the notion of walk. In short, the existence of a temporal walk from node v_i to v_j implies that we can travel from v_i to v_j possibly through other nodes by traversing events. Differently from the case of static networks, the time flows while a temporal walk is occurring. A walk has to wait at intermediary nodes, inducing waiting times, before an event appears to enable the walker to move to a neighbouring node. A temporal walk is a causal walk and not allowed to use events that have occurred in the past. This causality constraint often makes temporal walks very different from walks in the corresponding aggregate network. Many concepts and analysis tools for temporal networks, such as distance, connectivity and distance-based centrality measures, are built upon temporal walks, as we will see in the following. The temporal walk is also called the journey [Xuan *et al.* (2003)], schedule-conforming path [Berman (1996)], time-respecting path [Kempe *et al.* (2000); Holme (2005); Holme and Saramäki (2012)], temporal path [Tang *et al.* (2010b); Pan and

Saramäki (2011)] or non-decreasing path [Cheng *et al.* (2003)].

In the event-based representation, a temporal walk from node v_0 to node v_n is defined by a sequence of events

$$\{(v_0, v_1, t_1), (v_1, v_2, t_2), \ldots, (v_{n-1}, v_n, t_n)\}, \tag{4.5}$$

where $t_i \leq t_{i+1}$ ($1 \leq i \leq n-1$). Although t_i has represented the ith event in the event-based representation of a temporal network (Eq. (4.2)), here we use t_i to denote the time of the ith event on a temporal walk to avoid abuse of notation. The definition for the snapshot representation is the same; we interpret (v_{i-1}, v_i, t_i) as a link (v_{i-1}, v_i) in snapshot $G(t_i)$. The departure and arrival times of the temporal walk are defined as t_1 and t_n, respectively. The length of a temporal walk is defined by either its number of hops, n, or its duration, $t_n - t_1$.

Similarly to the definition of walk in static networks, we allow a temporal walk to visit the same node more than once. Likewise, we say that a temporal walk is a temporal path when $v_0, \ldots, v_n$ are different. Previous literature is not necessarily strict between the distinction between the temporal walk and path and often uses the term temporal path in the meaning of the temporal walk.

If we disregard the temporal information (i.e., aggregate network), we always overcount the number of walks and paths. This is because, in the aggregate network, walks unrealisable in the temporal network are also counted as walks. For example, in the temporal network shown in Fig. 4.2(a), there are three temporal walks from v_1 to v_4, i.e., $\{(v_1, v_2, t_1), (v_2, v_3, t_3), (v_3, v_4, t_5)\}$, $\{(v_1, v_2, t_1), (v_2, v_4, t_6)\}$ and $\{(v_1, v_3, t_4), (v_3, v_4, t_5)\}$. In contrast, there are infinitely many walks from v_1 to v_4 in the static, aggregate network shown in Fig. 4.2(b). In addition, there are less temporal paths than paths in the corresponding aggregate network. Because all temporal walks in Fig. 4.2(a) are temporal paths, there are three temporal paths. In contrast, Fig. 4.2(b) allows four paths succinctly represented as $v_1 \rightarrow v_2 \rightarrow v_3 \rightarrow v_4$, $v_1 \rightarrow v_2 \rightarrow v_4$, $v_1 \rightarrow v_3 \rightarrow v_2 \rightarrow v_4$ and $v_1 \rightarrow v_3 \rightarrow v_4$. If we distinguish the two duplicated links (v_3, v_4) as different links, there are even more paths in the aggregate network.

The temporal walk is not a symmetric notion even for an undirected temporal network. For example, consider the situation in which we have observed the first four events shown in Fig. 4.2(a). At this moment, a temporal walk from v_1 to v_4 does not exist. However, a temporal walk from v_4 to v_1 does exist, i.e., $\{(v_4, v_3, t_2), (v_3, v_1, t_4)\}$.

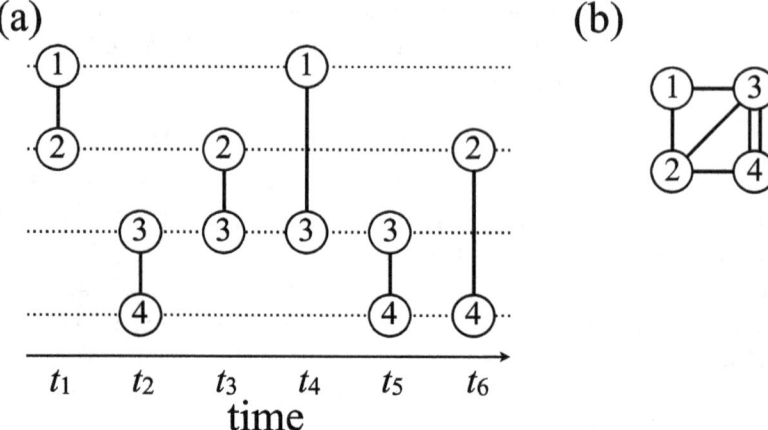

Fig. 4.2   (a) A temporal network on $N = 4$ nodes. (b) Corresponding aggregate network, where link (v_3, v_4) is explicitly duplicated.

We say that a node v_j is reachable from v_i or v_i is temporally connected to v_j if there is a temporal walk from v_i to v_j. Reachability is not a symmetric relationship; v_i may not be reachable from v_j even if v_j is reachable from v_i. Reachability is not a transitive relationship, either. In other words, even if v_1 is temporally connected to v_2 and v_2 is temporally connected to v_3, v_1 may not be temporally connected to v_3 (Fig. 4.3). In contrast, in static networks, either undirected or directed, if a (directed) path exists from v_1 to v_2 and another from v_2 to v_3, a (directed) path exists from v_1 and v_3, hence the relationship is transitive. Finally, reachability is time-dependent. Node v_2 reachable from v_1 in a time window does not guarantee the same reachability in a different time window.

4.2.2   Temporal distances

A broad range of metrics defined by walks and paths in static networks is generalised to temporal networks by using temporal walks and paths. We look at them in this and the following sections.

There exist at least three distinct measures of distance in temporal networks [Xuan et al. (2003)], each leading to its own version of the temporal walk with the smallest distance. When the temporal distance is discussed, we focus on temporal paths without loss of generality. This is because, if a smallest-distance walk visits the same node (e.g., v) more than once, we

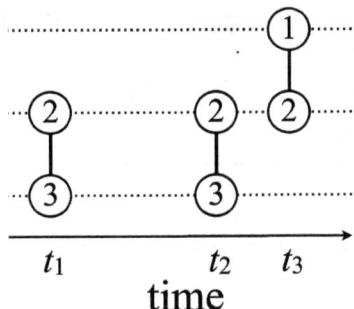

Fig. 4.3 A temporal network on $N = 3$ nodes. v_1 is temporally connected to v_2, v_2 to v_3, but not v_1 to v_3.

can omit the redundant segment of the walk (i.e., the closed walk starting from the first visit to v and ending at the second visit to v) to make the distance smaller or unchanged. Note that the situation here is slightly different from that of static networks, for which a smallest-distance walk is always a path.

Denote a temporal path from node u to node v by $\{(u, v_1, t_1), (v_1, v_2, t_2), \ldots, (v_{n-1}, v, t_n)\}$.

(1) The shortest distance from u to v at time t is defined by

$$d_{\text{short}}(u, v; t) = \min\{n : t_1 \geq t\}. \tag{4.6}$$

It is equal to the minimum number of hops necessary to travel from u to v along temporal paths. The temporal paths are constrained to start after time t. This is a topological distance measure.

(2) The foremost distance from u to v at time t is defined by

$$d_{\text{fore}}(u, v; t) = \min\{t_n - t : t_1 \geq t\}. \tag{4.7}$$

It is equal to the minimum amount of time necessary to travel from u to v after time t. The length is measured in the time domain.

(3) The travel-time distance, called the fastest distance in [Xuan *et al.* (2003)], from u to v at time t is defined by

$$d_{\text{travel}}(u, v; t) = \min\{t_n - t_1 : t_1 \geq t\}. \tag{4.8}$$

In contrast to d_{fore}, the travel-time distance favours the temporal path whose duration is as small as possible, independently of the initial waiting period $t_1 - t$ before the temporal path actually begins. This temporal distance is also measured in the time domain.

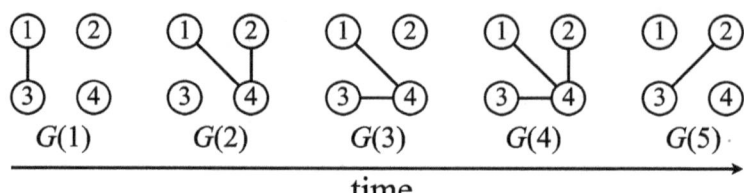

Fig. 4.4   A temporal network in the snapshot representation.

Different definitions of the temporal distance are appropriate in different situations. For instance, in the case of public transportation networks, where links correspond to connections between stations in a city, d_{short} will be preferred if a passenger aims at minimising the number of connections. In contrast, one should use d_{fore} if the goal is to arrive at the destination as early as possible. One should use d_{travel} to make the travel time as small as possible. It should be noted that the symmetry is broken for any of the three distance measures even if the events are undirected. It should also be noted that d_{short} is dimensionless, while d_{fore} and d_{travel} have the dimension of time.

Equations (4.6), (4.7) and (4.8) define time-dependent distance measures, where argument t is the time after which temporal paths are allowed to start. The distance between two nodes vary over time in general for many reasons, e.g., diurnal and weekly cycles for human communication networks.

The three definitions of the temporal distance are applicable to both event-based and snapshot representations of temporal networks. However, we have to be careful when multiple events may occur on different links at the same time, which is customary in particular in the snapshot representation. In this situation, we have to specify the number of hops allowed at each time point. By definition, the temporal path given by Eq. (4.5) allows one to make any number of hops at given t because $\{t_1, \ldots, t_k\}$ is constrained by $t_j \le t_{j+1}$, not $t_j < t_{j+1}$. In the snapshot representation, this assumption implies that one can instantly travel from node u to v if they are connected in a snapshot. For example, $d_{\text{fore}}(v_1, v_2, t = 1) = 2$ in the snapshot representation of a temporal network shown in Fig. 4.4, because nodes 1 and 2 are connected in snapshot G_2.

However, this assumption may be unrealistic because making many hops at time t implies infinitely, or at least very, fast movement. This is particularly so in the event-based representation in which the time is often regarded to be continuous. In fact, in many previous literature, only a single step is implicitly allowed at a certain time. This assumption corresponds to a stronger condition $t_j < t_{j+1}$ imposed on the temporal path. With this modification, $d_{\text{short}}(u, v; t)$, $d_{\text{fore}}(u, v; t)$ and $d_{\text{travel}}(u, v; t)$ increase (at least do not decrease) as compared to the case $t_j \leq t_{j+1}$. For example, in Fig. 4.4, the foremost distance with this altered definition increases to $d_{\text{fore}}(v_1, v_2, t = 1) = 4$. An intermediate assumption is to allow at most h hops at the same time t, or within a snapshot in the snapshot representation, where h, called horizon, is determined beforehand [Tang *et al.* (2009, 2010a); Takaguchi *ct al.* (2012)]. Different values of h define different temporal paths and temporal distance measures. The cases $h = 1$ and $h \to \infty$ correspond to the restrictions $t_j < t_{j+1}$ and $t_j \leq t_{j+1}$, respectively. A plausible choice of h depends on the time resolution of temporal network data, particularly in the case of the snapshot representation, and on the physics of the diffusive process underlying the definition of the temporal walk and distance.

Corresponding to each notion of temporal distance, we can define the average distance (i.e., average over node pairs) and diameter of temporal networks by Eqs. (3.13) and (3.14), respectively. The average distance and diameter are time-dependent measures because Eqs. (4.6), (4.7) and (4.8) are time-dependent.

We are often interested in characterising the distance between two nodes in a time-independent manner. There are at least three ways to derive a time-independent distance measure from a time-dependent one. First, for the shortest and travel-time distances, it is intuitive to use the smallest value across time [Xuan *et al.* (2003)]. In other words, $\min_t d_{\text{short}}(u, v; t)$ is the smallest number of hops necessary to travel from u to v anytime. $\min_t d_{\text{travel}}(u, v; t)$ is the smallest time necessary to travel from u to v.

Second, a time-independent distance measure is defined if we force $t = 0$. For example, $d_{\text{fore}}(u, v; 0)$ is equal to the earliest arrival time at v starting from u [Xuan *et al.* (2003)].

Third, a time-dependent distance measure averaged over the entire observation period defines a time-independent distance measure. However, averaging the distance across time is not an easy exercise, and certain choices have to be made [Pan and Saramäki (2011)]. A complication comes from the finiteness of the observation period, $t_{\max}$. The total number of future connections decreases as time increases towards $t_{\max}$. Therefore, the likelihood that a temporal path exists between any pair of nodes decreases with time. In particular, $d(v, v'; t_{\max}) = \infty$ for any $v \neq v'$. Let us describe two ways to circumvent the problem. First, we may consider early times when the temporal distance is finite. The problem with this approach is that it introduces a bias in favour of temporal paths taking place early within the observation period. Second, we may introduce a periodic boundary condition in time to identity $t = t_{\max}$ with $t = 0$. Then, the temporal distance is always finite as long as the corresponding aggregate network is connected. The problem of the periodic boundary condition is that it introduces artificial temporal paths between node pairs, which do not exist in the given data. By using the periodic boundary condition in a restricted manner, we can mitigate the effect of artificial temporal paths [Pan and Saramäki (2011)].

4.2.3 *Vector clock*

By applying the concept of the foremost distance measure, we can quantify and calculate the recency of information that a node obtains about another node. By looking backward in time, the so-called temporal view $\phi(v, u, t)$ that node v has about node u at time t is defined as the latest time $t'(\leq t)$ such that a temporal path departing node u at t' can arrive at v before time t [Kossinets *et al.* (2008)]. By definition, $\phi(v, v, t) = t$. The event through which v gains the information about u as of time t' does not have to be an event on link (u, v). It may come indirectly through a third node v' via two events (u, v', t') and (v', v, t''), where $t' < t'' < t$. For example, at $t = t_3$ in the temporal network in the event-based representation shown in Fig. 4.5(a), node v_1 has obtained the most recent information about node v_2 not directly from v_2 but indirectly via node v_3 using two events (v_2, v_3, t_2) and (v_1, v_3, t_3).

The information latency is defined by $t - \phi(v, u, t)$ and represents the time that has passed since v obtained the most up-to-date information about u. For the temporal network shown in Fig. 4.5(a), the temporal view and information latency that v_1 has about v_2 are shown in Fig. 4.5(b). The vector clock for node v_i is defined by $\phi(v_i, t) = (\phi(v_i, v_1, t), \ldots, \phi(v_i, v_N, t))$ and represents the list of most up-to-date information that v_i has about all nodes at time t [Lamport (1978); Mattern (1989); Kossinets *et al.* (2008)].

The vector clocks for all nodes v can be simultaneously calculated in linear time in terms of the number of events in the temporal network. To do this, we need to sweep the list of events in ascending order of time just once. First, we initialise the vector clock by assigning a null symbol to all $\phi(v, u, 0)$

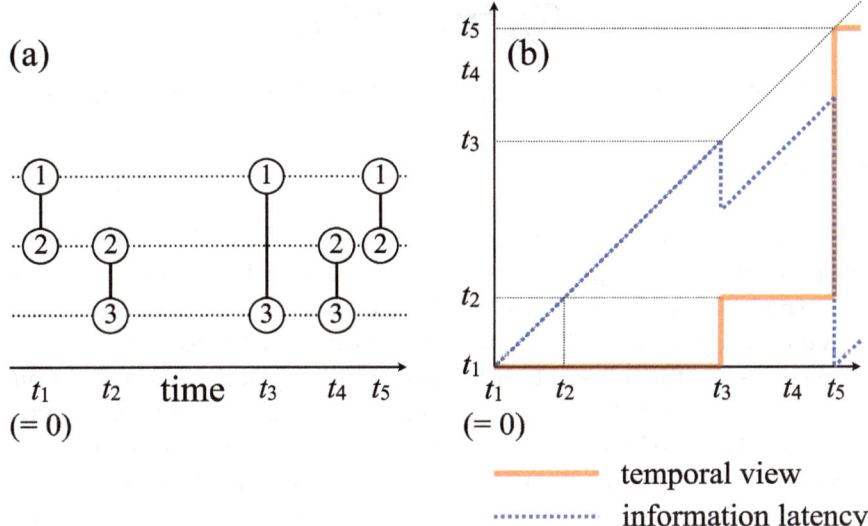

Fig. 4.5 (a) A temporal network on $N = 3$ nodes. (b) Temporal view $\phi(v_1, v_2, t)$ and the information latency $t - \phi(v_1, v_2, t)$, quantifying the novelty of the information that v_1 has about v_2.

$(u \neq v)$. This is because v does not have any information about u before any event occurs; u and v have not been connected by a temporal path. We also initialise $\phi(v, v, 0) = 0$ $(1 \leq v \leq N)$. Second, when the ith event (u_i, v_i, t_i) arrives, we set vector clock $\phi(v_i, t_i)$ to be the coordinate-wise maximum of the most recent vector clocks $\phi(u_i, t_{i'})$ and $\phi(v_i, t_{i''})$, where $t_{i'}$ and $t_{i''}$ $(1 \leq i', i'' < i)$ are the times of the most recent vector clock for u_i and v_i, respectively. Third, we set $\phi(v_i, v_i, t_i) = \phi(v_i, u_i, t_i) = t_i$. If the event (u_i, v_i, t_i) is undirected, as assumed in most of the present book, we substitute the updated $\phi(v_i, t_i)$ in $\phi(u_i, t_i)$. Then, we process the next event $(u_{i+1}, v_{i+1}, t_{i+1})$ in the same manner.

The advance in v's vector clock $\phi(v, t)$ is defined as the difference between the component-wise sum of $\phi(v, t)$ right after the event (u, v, t) and that before the event [Kossinets *et al.* (2008)]. The advance quantifies the amount of information that event (u, v, t) brings to node v. An event with a large advance value is considered to be valuable. Empirically, the advance of an event (u, v, t) tends to be large if the removal of the link (u, v) enlarges the distance between u and v in the aggregate network.

Fig. 4.6   Temporal connectedness. An arrow indicates that a node is reachable from another.

4.3   Components

In temporal networks, the constraint that a temporal path is a time-ordered sequence leads to unexpected properties of connected components. As we have seen in the definition of reachability (Section 4.2.2), connectivity is not a symmetric relation even if the original network is undirected, as in the case of static directed networks. In other words, v_1 may not be reachable from v_2 even if v_2 is reachable from v_1 in undirected temporal networks. Therefore, the concept of component of both undirected and directed temporal networks is developed based on that for static directed networks as follows [Nicosia *et al.* (2012)] (equivalent to open strongly connected components in [Bhadra and Ferreira (2003)]).

For simplicity, let us focus on undirected temporal networks. Nodes v_i and v_j in undirected temporal networks are defined to be temporally connected if v_i and v_j are mutually reachable [Kempe *et al.* (2000); Nicosia *et al.* (2012)]. For example, in a three-node temporal network shown in Fig. 4.6, where an arrow indicates the reachability relationship, nodes v_1 and v_2 are temporally connected, and so are v_2 and v_3.

With this relationship, we can define the temporal connected component. However, formulation and computation of components are much more involved for temporal than static networks. This is because reachability in temporal networks is not a transitive relationship (Section 4.2.2). A consequence is that temporal connectedness is not a transitive relationship, either. Therefore, for temporal networks, it is impossible to define the notion of connected component purely from the definition of connectedness for pairs of nodes. If the network shown in Fig. 4.6 were a static directed network, the three nodes constitute a single strongly (and weakly) connected component. This is because not only node pairs (v_1, v_2) and (v_2, v_3) but also (v_1, v_3) are bidirectionally connected. However, because Fig. 4.6 represents a reachability relationship in a given temporal network, it does not imply that v_1 and v_3 are even unidirectionally connected. Then, we cannot uniquely determine which of v_1 or v_3 forms a temporal connected component with v_2.

The lack of transitivity in the reachability relationship is not the only obstacle against the temporal connected component that we are going to define. Mutual reachability between v_1 and v_2 does not imply that there is a temporal walk $v_1 \rightarrow v_2 \rightarrow v_1$. This is in a stark contrast with the case of static networks. In addition, the time dependence of reachability implies that the temporal connected component will also be a time-dependent concept.

With these caveats in mind, the temporal connected component is defined as a maximal set of nodes in which each node pair in the set is temporally connected. For directed temporal networks, the strong and weak connectedness (and connected components) are defined by looking at reachability between pairs of nodes for the original directed temporal network and the undirected version of the temporal network, respectively. Even with this definition, the aforementioned problem has not been resolved. In other words, a node may belong to multiple connected components, which never happens in static networks. For the same reason, temporal networks allow for non-unique partitioning into connected components, whereas partitioning is unique for static networks.

Finding temporal connected components is also a computationally demanding problem. In static networks, the connectedness can be checked by sweeping the link list once, such that the computational time is $O(M)$. In contrast, in temporal networks, we have to check the temporal connectedness for every pair of nodes individually, making the computational time $O(N^2)$ [Nicosia *et al.* (2012)]. Because $M \ll N^2$ for sparse networks, the computation of temporal connected components is relatively slow for sparse networks. Furthermore, finding all connected components in temporal networks is mapped to an NP-complete problem (and hard to approximate). Therefore, it is difficult to calculate them for large networks.

To understand both conceptual and computational difficulties of temporal components, it is instructive to use the associated affine graph. The associated affine graph for a given temporal network is a static undirected network such that a link is laid if and only if the two nodes are temporally strongly connected. Finding a strongly connected component in a temporal network is equivalent to finding a maximal clique (i.e., a clique that is not included in any larger clique), which is known to be an NP-complete problem. In addition, as the example in Fig. 4.7 shows, the partitioning of the network, i.e., partitioning into maximal cliques, is not unique. Disregarding the case of trivial, single-node cliques, the associated affine graph allows two partitions, i.e., $\{1, 2, 3\} \cup \{4, 5\}$ and $\{2, 4, 5\} \cup \{1, 3\}$. Figure 4.7 also

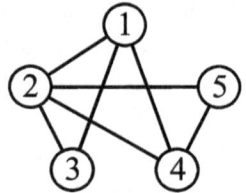

Fig. 4.7   An associated affine graph. Example from [Nicosia *et al.* (2012)].

implies that a node can belong to different strongly connected components even in the case of undirected temporal networks, similar to overlapping communities.

4.4   Temporal coherence of a triangle

The clustering coefficient for static networks is a normalised count of triangles in the network (Section 3.4). The rationale behind this idea is that three nodes forming a triangle is the smallest group (of size three) beyond a single link (which is a group of size two). In a social network analogy, a triangle would represent friendship or a group of workmates composed of three persons.

In temporal networks, the notion of triangle is not straightforward. Consider two temporal networks in the snapshot representation shown in Fig. 4.8. Both networks are composed of $N = 3$ nodes, $t_{\max} = 8$ snapshots, and yield the same aggregate network. In particular, the three nodes form a triangle from the viewpoint of the static network. However, it is not necessarily the case in temporal networks. In Fig. 4.8(a), although each pair of nodes communicates via events, there is no case in which three nodes simultaneously communicate. For example, when nodes v_1 and v_3 are connected (i.e., $t = 3$, 7, and 8), node v_2 is isolated. Node v_2 is connected to v_1 and v_3 but in different times. Therefore, although each pair of nodes is a friend pair, the three nodes as a whole may not be a social group. In contrast, in Fig. 4.8(b), it is often the case that three nodes form triangles (i.e., $t = 2$ and 7). Therefore, it is likely that the three individuals really form a social group.

To distinguish between the two cases, Eckmann introduced a mutual information measure, called the temporal coherence of a triangle, denoted by I_T [Eckmann *et al.* (2004)]. It is defined for a triplet of nodes in directed temporal networks in the snapshot representation and given for the triplet

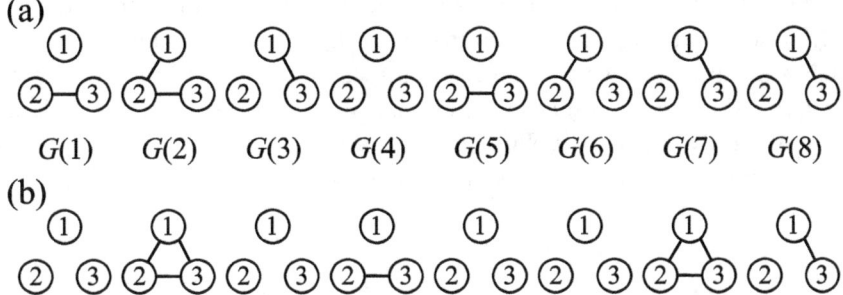

Fig. 4.8   Two temporal networks yielding the same intensity of a triangle in the static network. (a) The triangle is temporally incoherent. (b) The triangle is temporally coherent. $N = 3$. $t_{\max} = 8$.

of the ith, jth and kth nodes by

$$I_T = \sum_{b_{ij},b_{ji},b_{ik},b_{ki},b_{jk},b_{kj}\in\{0,1\}} p_{ijk}(b_{ij},b_{ji},b_{ik},b_{ki},b_{jk},b_{kj})$$

$$\times \log \frac{p_{ijk}(b_{ij},b_{ji},b_{ik},b_{ki},b_{jk},b_{kj})}{p_{ij}(b_{ij},b_{ji})p_{ik}(b_{ik},b_{ki})p_{jk}(b_{jk},b_{kj})}. \tag{4.9}$$

In Eq. (4.9), $p_{ij}(b_{ij},b_{ji}) = N_{ij}(b_{ij},b_{ji})/t_{\max}$, for example, where $N_{ij}(b_{ij},b_{ji})$ is the number of snapshots in which $A_{ij}(t) = b_{ij}$ and $A_{ji}(t) = b_{ji}$. Therefore, $p_{ij}(1,1)$ is the fraction of snapshots in which v_i and v_j are mutually connected, $p_{ij}(0,1)$ is the fraction of snapshots in which the directed link from v_j to v_i is present but not vice versa, and so forth. $p_{ijk}(b_{ij},b_{ji},b_{ik},b_{ki},b_{jk},b_{kj})$ is the fraction of snapshots in which $A_{ij}(t) = b_{ij}$ and so on. For example $p_{ijk}(1,1,1,1,1,1)$ is the fraction of snapshots in which the bidirectionally connected triangle is formed, and $p_{ijk}(0,0,0,0,0,0)$ is the fraction of snapshots in which no link is present.

I_T is the mutual information measuring the difference between the actual probabilities of the occurrence of the 2^6 events $p_{ijk}(b_{ij},b_{ji},b_{ik},b_{ki},b_{jk},b_{kj})$ and the null-case probabilities in which different (undirected) links are present or absent independently of each other, i.e., $p_{ij}(b_{ij},b_{ji})p_{ik}(b_{ik},b_{ki})p_{jk}(b_{jk},b_{kj})$. If the three undirected links are activated independently, we would have $I_T = 0$. If these links tend to be synchronised, we have a large value of I_T. The maximum of I_T, which can be used as a normalisation constant, is equal to $\log 16$, which is realised when $p_{ijk}(0,0,0,0,0,0) = p_{ijk}(0,1,0,1,0,1) = p_{ijk}(1,0,1,0,1,0) = p_{ijk}(1,1,1,1,1,1) = 1/4$. In this case, we obtain $p_{ij}(b_{ij},b_{ji}) = p_{ik}(b_{ik},b_{ki}) = p_{jk}(b_{jk},b_{kj}) = 1/4$ for all $b_{ij},b_{ji},b_{ik},b_{ki},b_{jk},b_{kj} \in \{0,1\}$.

If we threshold triplets of nodes by a value of I_T, we can extract temporally coherent triangles. For example, with $I_T = 0.5$, an email traffic data set has much less coherent triangles than the number of triangles in static networks. This result implies that many triangles in the static network are not temporally coherent [Eckmann *et al.* (2004)].

In undirected temporal networks, the normalised variant of Eq. (4.9) is given by

$$I_T' = \frac{1}{\log 4} \sum_{b_{ij}, b_{ik}, b_{jk} \in \{0,1\}} p_{ijk}(b_{ij}, b_{ik}, b_{jk}) \log \frac{p_{ijk}(b_{ij}, b_{ik}, b_{jk})}{p_{ij}(b_{ij})p_{ik}(b_{ik})p_{jk}(b_{jk})},$$

(4.10)

where $p_{ij}(0)$ and $p_{ij}(1)$, for example, are the fraction of snapshots in which link (i,j) is absent and present, respectively. In Fig. 4.8(a), $p_{12}(0) = 3/4$ and $p_{12}(1) = 1/4$. $p_{ijk}(b_{ij}, b_{ik}, b_{jk})$ is the fraction of snapshots in which the triplet state (b_{ij}, b_{ik}, b_{jk}) is observed. In Fig. 4.8(b), the fraction of snapshots in which no link is present is equal to $p_{123}(0,0,0) = 1/2$. The fraction of snapshots in which a connected triangle forms is equal to $p_{123}(1,1,1) = 1/4$. The maximum of the mutual information on the right-hand side of Eq. (4.10), excluding the factor $1/\log 4$, is equal to $\log 4$. This value is realised by $p_{ijk}(0,0,0) = p_{ijk}(1,1,1) = 1/2$, which yields $p_{ij}(0) = p_{ij}(1) = p_{ik}(0) = p_{ik}(1) = p_{jk}(0) = p_{jk}(1) = 1/2$. Therefore, $0 \leq I_T' \leq 1$. For the temporal networks shown in Figs. 4.8(a) and 4.8(b), we obtain $I_T' \approx 0.282$ and $I_T' \approx 0.485$, respectively. This result confirms that a temporal network with more synchronised (i.e., coherent) triangles would yield a larger value of I_T'.

4.5   Centrality

Temporal networks are dynamical objects. Similarly to the case of the temporal distance, we distinguish two types of node centrality for temporal networks: time-independent and time-dependent centrality measures.

A time-independent centrality for temporal networks is defined as a summary over time. It quantifies the overall importance of a node during the observation period. In contrast, a time-dependent centrality intends to capture temporal changes in the importance of a node, to model the possibility that a node important at a certain time may not be important at other times. Consider the weighted degree (also called the node's strength; here we simply call it the degree) as a simple centrality measure for static networks. The time-independent variant of the degree centrality for node v_i in a temporal network would be defined as the sum of the instantaneous degree of v_i across time. This quantity is equal to the total number of events involving v_i in the case of the event-based representation and the sum of the degree over all snapshots in the case of the snapshot representation. This centrality coincides with the degree in the aggregate network. A simple time-dependent variant of the degree centrality would be defined by the instantaneous degree of v_i. Then, the centrality of each node is a time series of the instantaneous degree.

In this section, we present some centrality measures for temporal networks belonging to each category. It should be noted that nodes that are central in static networks are not necessarily central in the corresponding temporal networks and vice versa.

4.5.1 *Time-independent centrality*

There are centrality measures for static networks based on the concept of distance. They can be extended to time-independent centrality measures for temporal networks. The idea is to replace the distance measure appearing in the definition of the centrality measure by a time-independent distance measure for temporal networks, such as $\min_t d_{\text{short}}(u, v; t)$. A different measure of time-independent temporal distance defines a different centrality measure for temporal networks.

4.5.1.1 *Temporal closeness*

The temporal closeness is equal to the reciprocal of the mean of the temporal distance between a node and any other node [Tang *et al.* (2010a)]. A definition is given by Eq. (3.45) with $d(v_i, v_j)$ replaced by a time-independent temporal distance. As in other distance-based centrality measures for temporal networks, the temporal closeness centrality has the shortest, foremost and travel-time versions.

Similarly to the case of the closeness centrality for static networks, the centrality of node v_i would be equal to zero if v_i is not temporally connected to at least one node. As compared to static network data, this situation is likely to occur more frequently in temporal network data, which may be short or have some relatively isolated nodes [Holme and Saramäki (2012)]. Therefore, a better measure for closeness would be to use the temporal efficiency [Tang *et al.* (2009, 2010b)], i.e.,

$$\frac{1}{N-1} \sum_{j=1; j\neq i}^{N} \frac{1}{d(v_i, v_j)}. \tag{4.11}$$

With this definition, a pair of unconnected nodes yields a null contribution to a single term of Eq. (4.11) without affecting other terms. Some literature refers to the quantity given by Eq. (4.11) as temporal closeness centrality [Pan and Saramäki (2011)].

It should be noted that, for each definition, the centrality value changes even for undirected temporal networks if we exchange $d(v_i, v_j)$ to $d(v_j, v_i)$. This is not the case for static networks. There are also other definitions of temporal closeness [Holme (2015)].

4.5.1.2 *Temporal betweenness*

Tang and colleagues defined the temporal betweenness centrality as follows [Tang *et al.* (2010a)]. Consider the snapshot representation of a temporal network. Then, adopt one of the definitions of temporal distance to determine how to measure the length of temporal walks (e.g., by the duration of the walk, $t_n - t_1$). The temporal betweenness centrality of the ith node at time t is defined by

$$\text{betweenness}_i(t) = \frac{1}{(N-1)(N-2)} \sum_{j=1; j\neq i}^{N} \sum_{j'=1; j'\neq i, j}^{N} \frac{\sigma_{jj'}^i(t)}{\sigma_{jj'}}. \tag{4.12}$$

In fact, betweenness$_i(t)$ is a time-dependent centrality measure, which is the topic of Section 4.5.2. In Eq. (4.12), $\sigma_{jj'}$ is the number of the minimum-distance temporal path from the jth to the j'th nodes in the entire observation period $[0, t_{\max}]$. $\sigma_{jj'}^i(t)$ is the number of the minimum-distance temporal path from the jth to the j'th nodes that passes through the ith node and stays there at time t (i.e., reaches the ith node at time t or earlier and does not move to another node before t). The fraction of the minimum-distance paths that pass through the ith node is averaged over all starting and ending nodes of such paths. This normalisation is the same as that for the static-network counterpart (Eq. (3.46)). It should be noted that, even for undirected temporal networks, the paths from one node to another and those in the opposite direction are separately considered in Eq. (4.12) because the number and the distance of the qualified paths are generally different between the two directions.

A time-independent temporal betweenness centrality is given by

$$\text{betweenness}_i = \frac{1}{t_{\max}} \sum_{t=1}^{t_{\max}} \text{betweenness}_i(t). \qquad (4.13)$$

With this definition, the effect of a minimum-distance temporal path that stays at the ith node before transiting to another node is emphasised. An alternative definition is

$$\text{betweenness}'_i = \frac{1}{(N-1)(N-2)} \sum_{j=1;j\neq i}^{N} \sum_{j'=1;j'\neq i,j}^{N} \frac{\sigma^i_{jj'}}{\sigma_{jj'}}, \qquad (4.14)$$

where $\sigma^i_{jj'}$ is the number of the minimum-distance temporal paths from the jth to the j'th nodes that pass through the ith node.

As Eq. (4.14) shows, there seems to be several possibilities to extend the betweenness centrality for static networks to temporal networks in natural ways. In addition, these definitions vary upon the choice of the temporal distance measure. The original definition of betweenness$_i$ employs the foremost distance. However, we should probably use the shortest distance (i.e., number of hops), not the foremost or travel-time distance, to define the temporal betweenness centrality if we adhere with the event-time representation. This is because, due to the continuous-time nature of the event-time representation, there would not be multiple minimum-distance paths between a given node pair with the foremost or travel-time distance. This is in particular the case when the time resolution of the temporal network is high. Then, the fraction of the minimum-distance paths passing through v_i would be equal to zero or one for almost all pairs of nodes, which deviates from the concept of betweenness.

4.5.1.3 *Dynamical communicability*

The dynamic communicability extends the idea of Katz centrality (Section 3.7) to temporal networks [Grindrod *et al.* (2011)]. Consider temporal walks on a temporal network in the snapshot representation. We allow the walker to traverse an arbitrary number of links within a snapshot. The dynamic communicability matrix uses a scaled number of walks between nodes calculated as follows:

$$Q_{\text{dc}} \equiv [I - \alpha A(1)]^{-1} [I - \alpha A(2)]^{-1} \cdots [I - \alpha A(t_{\max})]^{-1}, \qquad (4.15)$$

which generalises Eq. (3.47). In Eq. (4.15), a temporal walk of length ℓ is discounted by a factor of α^ℓ. For example, both a temporal walk of length three using snapshots $A(1)$, $A(1)$, and $A(3)$ and another temporal path using $A(1)$, $A(2)$, and $A(3)$ contribute to the dynamic communicability matrix by α^3. In the former temporal walk, a contribution of $\alpha^2 [A(1)]^2$ originates from the Taylor expansion of $[I - \alpha A(1)]^{-1}$ on the right-hand side of Eq. (4.15) and $\alpha A(3)$ from $[I - \alpha A(3)]^{-1}$. All $[I - \alpha A(t)]^{-1}$ ($t \neq 1, 3$) contribute I. For the matrix inverses appearing in Eq. (4.15) to exist, α must be smaller than $\min_{1 \leq t \leq t_{\max}}$ [largest eigenvalue of $A(t)]^{-1}$.

Similarly to the derivation of the Katz centrality, matrix Q_{dc} itself is not a node centrality. By summing over all the destination nodes of the walks, Grindrod and colleagues defined the broadcast centrality for node v_i. Precisely, the broadcast centrality is equal to $\sum_{j=1}^{N}(Q_{\mathrm{dc}})_{ij}$ up to normalisation. It quantifies the importance of v_i in sending information to all nodes using temporal walks. The opposite concept, the receive centrality is defined for v_i by $\sum_{j=1}^{N}(Q_{\mathrm{dc}})_{ji}$. Even if each snapshot is an undirected network, the broadcast and receive centrality values for the same node are different from each other in general because Q_{dc} is an asymmetric matrix.

On an email communication data set, the dynamic communicability is correlated with the impact of individual nodes on epidemic spreading in a stochastic epidemic process model [Mantzaris and Higham (2013)].

4.5.1.4   *TempoRank*

The PageRank is defined as the stationary density of a discrete-time random walk on the network (Section 3.7.4). Because the random walk can be extended to the case of temporal networks, so can the PageRank. TempoRank is one such implementation for temporal networks in the snapshot representation [Rocha and Masuda (2014)]. Although the original definition accommodates link weights in snapshots, here we state the case of unweighted snapshots for simplicity.

If each node is adjacent to at most one node in each snapshot and a walker is forced to move unless the currently visited node is isolated, the walk is deterministic, not random. For example, in the temporal network on $N = 4$ nodes shown in Fig. 4.9, the walker starting from node v_1 always goes to node v_4 when the temporal network terminates at $t_{\max} = 3$. To make the walk random, we assume that, in each snapshot, a walker stays at the currently visited node with some probability even if it can move to a neighbour. For example, in the first snapshot in Fig. 4.9, the walker at v_1 does not move with probability q and moves to v_2 with probability $1 - q$, where $0 < q < 1$. Then, the walker starting from v_1 may be located at

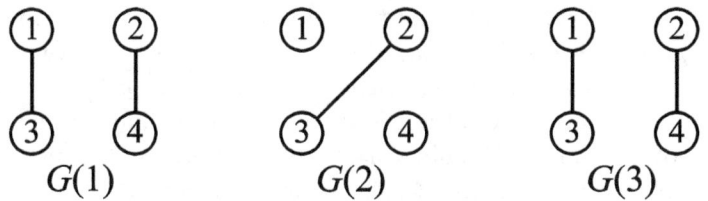

Fig. 4.9 A temporal network on $N = 4$ nodes and $t_{\max} = 3$ snapshots.

any of the four nodes at time $t_{\max}(= 3)$. This parameter, q, allows us to explore situations in which diffusion is slower than the time scale on which snapshots are observed. Therefore, it can be seen as a complementary to the horizon h (Section 4.2.2).

In general, we define the transition probability from v_i to v_j in snapshot $A(t)$ as

$$
T_{ij}(t) = \begin{cases} \delta_{ij} & (k_i(t) = 0, 1 \leq j \leq N), \\ q^{k_i(t)} & (k_i(t) \geq 1, i = j), \\ (A(t))_{ij}(1 - q^{k_i(t)})/k_i(t) & (k_i(t) \geq 1, i \neq j), \end{cases} \qquad (4.16)
$$

where $k_i(t) = \sum_{j=1}^{N}(A(t))_{ij}$ is the degree of v_i at time t. Note that $\sum_{j=1}^{N} T_{ij}(t) = 1$. Equation (4.16) implies that the walker does not move with probability $q^{k_i(t)}$ and otherwise move to a neighbour in $A(t)$ with the equal probability. The probability of not moving is set to $q^{k_i(t)}$ because each link present in the snapshot induces the move with probability $1 - q$ and the walker has to deny all them to remain at v_i. In this way, the walker's probability of not moving is not affected by whether multiple links are activated together in a snapshot or somewhat separately in time; such a difference would easily arise by changing the time resolution of temporal network data.

The transition matrix at time t is given by $T(t) = (T_{ij}(t))$. A one-cycle transition matrix for the temporal network is defined as

$$
T^{\mathrm{tp}} \equiv T(t_1)T(t_2)\cdots T(t_{\max}), \qquad (4.17)
$$

where tp stands for temporal. Under the periodic boundary condition, i.e., applying snapshot $A(1)$ after $A(t_{\max})$, the "stationary density" is given by the solution of the equation

$$
\boldsymbol{u}(1) = \boldsymbol{u}(1)T^{\mathrm{tp}}, \qquad (4.18)
$$

where $\boldsymbol{u}(1)$ is the N-dimensional row vector whose ith element is equal to the stationary density at v_i. However, the concept of stationary density should be treated with caveats for temporal networks. The stationary density in the usual sense does not exist in a given temporal network. This is because the network is ever changing. Even if the snapshots are applied periodically and indefinitely, the density at each node differs from snapshot to snapshot even in the equilibrium of the dynamics. $u(1)$ represents the equilibrium density when the density is observed just after $A(t_{\max})$ (i.e., just before $A(1)$) in every cycle. With this observation, the density converges to $u(1)$. The TempoRank is defined by the running average of the equilibrium density over the snapshots, i.e., $\overline{u} \equiv \sum_{t=1}^{t_{\max}} u(t)/t_{\max}$, where $u(t)$ is the stationary density when the observation is made just before applying $A(t)$.

4.5.2 *Time-dependent centrality*

We already saw some examples, explicit and implicit, of time-dependent centrality measures based on time-dependent distance measures. In this section, we explain a family of time-dependent centrality measures that do not use a temporal distance.

4.5.2.1 *Exponential discounting in time*

The importance of a node may fluctuate over time in a temporal network because the network itself varies over time. The easiest way to construct a time-dependent centrality measure for temporal networks in the snapshot representation is to define the centrality at time t as a static-network centrality measure calculated for snapshot $G(t)$. We already studied the degree as a pedagogical example of this case. However, unless the time resolution of the temporal network is low, snapshots are usually sparse matrices and perhaps disconnected networks. Then, the degree in each snapshot may not be a sensible choice because it is generally very small, perhaps equal to zero, in a majority of snapshots. Furthermore, snapshots are more often than not a highly fragmented network. Then, other centrality measures for the static network such as the betweenness centrality and the Katz centrality would not work because they are not very suitable for disconnected networks.

We can get around the problem by letting the impact of snapshots decay exponentially in time. In other words, we calculate

$$\tilde{A}(t) \equiv \sum_{t'=1}^{t} e^{-b(t-t')} A(t'), \qquad (4.19)$$

where $b > 0$ is the decay rate. Then, we calculate a centrality measure for the static network with adjacency matrix $\tilde{A}(t)$ to be used as the time-dependent centrality at time t. With this recipe, we can avoid the problem of sparse snapshots because $\tilde{A}(t)$, which takes into account all past snapshots, is not usually a sparse matrix. The same logic also functions for the event-based representation of temporal networks.

The exponential discounting also brings in another bonus. It gives an online updating rule for $\tilde{A}(t)$ as follows:

$$\tilde{A}(t+1) = e^{-b}\tilde{A}(t) + A(t+1). \tag{4.20}$$

Equation (4.20) implies that calculating $\tilde{A}$ at time $t+1$ does not require performing the full summation in Eq. (4.19) if $\tilde{A}$ has been calculated at the previous time step t. We obtain the variant of Eq. (4.20) for the event-based representation by replacing e^{-b} by $e^{-b(t_{n+1}-t_n)}$, where t_n is the time of the nth event in the entire event sequence of a given temporal network, after merging the event sequences on all links. When $b = \infty$, $\tilde{A}(t) = A(t)$ such that each snapshot is individually considered. When $b = 0$, $\tilde{A}(t) = \sum_{t'=1}^{t} A(t)$ such that $\tilde{A}(t)$ is essentially the aggregate network.

We have derived an updating rule for the adjacency matrix. In some cases, we can also obtain a similar updating rule for matrices or vectors used for calculating a time-dependent centrality measure. In this case, we can reuse the centrality values up to the previous time step to accelerate the computation and guide better understanding of the mechanism of the time-dependent centrality measure in question. We will explain two examples in the next sections. A price of this approach is that only centrality measures for static weighted networks can be used. This is because a matrix or vector that is updated online is inherently a weighted network even if the original snapshots (i.e., $A(t)$) are unweighted networks. In fact, we have observed that $\tilde{A}(t)$ is generally a weighted adjacency matrix. For example, the betweenness centrality is predominantly used for unweighted static networks although definitions for weighted networks exist. Therefore, it is probably not fruitful to apply this approach to the betweenness centrality.

4.5.2.2 *Running broadcast and receive centrality*

Broadcast and receive centrality measures have been extended to running broadcast and receive centrality measures [Grindrod and Higham (2013, 2014)]. The underlying design principle is that temporal walks that have started in a recent past are more important than those that have started earlier. The running dynamic communicability matrix is defined by the following update equation:

$$Q_{\text{dc-run}}(t_n) = \left[I + e^{-b(t_n - t_{n-1})} Q_{\text{dc-run}}(t_{n-1})\right] [I - \alpha A(t_n)]^{-1} - I, \quad (4.21)$$

where t_n is the time of the nth snapshot. The initial condition is given by $Q_{\text{dc-run}}(0) = 0$. In the usual snapshot representation of temporal networks, one snapshot arrives per unit time such that $t_n = n$. Here, we allow for general arrival times of the snapshot.

Equation (4.21) is interpreted as follows. Factor $[I - \alpha A(t_n)]^{-1}$ is the same as the dynamic communicability matrix (Eq. (4.15)); it enumerates walks on the static network represented by $A(t_n)$ with walks of length ℓ discounted by a factor of α^ℓ. Factor $\left[I + e^{-b(t_n - t_{n-1})} Q_{\text{dc-run}}(t_{n-1})\right]$ indicates that if a past snapshot is used in a temporal walk, the weight of the walk is suppressed by a factor of $e^{-b(t_n - t_{n-1})}$. For example, a walk whose first move occurs in snapshot $A(t_1)$ is counted in $Q_{\text{dc-run}}(t_1)$ (i.e., Eq. (4.21) with $n = 1$). Therefore, if we are at time t_n, the weight of this move is discounted by a factor of $e^{-bt_1} e^{-b(t_2 - t_1)} \cdots e^{-b(t_n - t_{n-1})} = e^{-bt_n}$ as well as a factor of α. Finally, $-I$ in the end of the right-hand side of Eq. (4.21) removes the null walk, cancelling with $I \times I$ resulting from the first term.

With $b = 0$, the non-running version of the dynamic communicability matrix is recovered in principle, i.e., $Q_{\text{dc}}(t_n) = I + Q_{\text{dc-run}}(t_n)$. The difference, I, corresponds to the fact that null walks are excluded in $Q_{\text{dc-run}}$ but not in Q_{dc}.

Finally, the running versions of the broadcast and receive centrality of node v_i are defined by $\sum_{j=1}^{N} (Q_{\text{dc-run}}(t_{\max}))_{ij}$ and $\sum_{j=1}^{N} (Q_{\text{dc-run}}(t_{\max}))_{ji}$, respectively.

4.5.2.3 *Dynamic win-lose score*

Node centrality has been used for ranking teams and players in sports. The win-lose score, originally proposed for ranking colleges in American football leagues as static networks [Park and Newman (2005)], was extended to account for temporal changes in the score of nodes [Motegi and Masuda (2012)]. The time-dependent centrality measure for temporal networks called the dynamic win-lose score is defined as follows.

Consider the snapshot representation where the time of the nth snapshot, t_n, is arbitrary, as in the case of the running broadcast and receive centrality. We define the snapshots based on the results of the matches. If node v_j defeats node v_i once at time t, we set $(A(t))_{ij} = 1$. Otherwise, $(A(t))_{ij} = 0$. In fact, multiple matches between the same or different node pairs are allowed in a single snapshot, in which case $A(t)$ is a weighted matrix. Matrix $A(t)$ is generally asymmetric.

Let us start with the case of static networks. The original win score of v_i, denoted by w_i, is defined by

$$
\begin{aligned}
w_i &= \sum_{j=1}^{N} A_{ji} + \alpha \sum_{j,j'=1}^{N} A_{j'j} A_{ji} + \alpha^2 \sum_{j,j',j''=1}^{N} A_{j''j'} A_{j'j} A_{ji} + \cdots \\
&= \sum_{j=1}^{N} \left(1 + \alpha \sum_{j'=1}^{N} A_{j'j} + \alpha^2 \sum_{j',j''=1}^{N} A_{j''j'} A_{j'j} + \cdots \right) A_{ji} \\
&= \sum_{j=1}^{N} (1 + \alpha w_j) A_{ji},
\end{aligned}
\tag{4.22}
$$

where A is the match matrix for the aggregate network [Park and Newman (2005)]. Equation (4.22) assumes that indirect wins (i.e., if v_1 defeats v_2 and v_2 defeats v_3, then v_1 is considered to indirectly defeat v_3) is discounted by α^ℓ, where ℓ is the length of the chain (e.g., $\ell = 2$ from v_1 to v_3). The direct wins are of course the most valued, but indirect wins also contribute to the win score. In particular, a node gains a large excess win score by beating a strong node because such a win would create many indirect wins with small ℓ. In the vector form, Eq. (4.22) is expressed as

$$
\boldsymbol{w} = \mathbf{1}A + \alpha \boldsymbol{w} A,
\tag{4.23}
$$

where $\boldsymbol{w} = (w_1 \cdots w_N)$ and $\mathbf{1} = (1 \cdots 1)$. Equivalently, the win score is given by $\boldsymbol{w} = \mathbf{1}A(I - \alpha A)^{-1}$. We can similarly define the lose score for each node by using $A^\top$ instead of A. The win-lose score is defined as the difference between the win score and the lose score. The win-lose score is well-defined only when α is smaller than the reciprocal of the largest eigenvalue of A.

For temporal networks, we posit that the effect of a win and loss decays exponentially in time at rate b. We define the dynamic win score at time

t_n through the dynamic win score matrix given by

$$
\begin{aligned}
W(t_n) \\
&= A(t_n) + e^{-b(t_n - t_{n-1})} \sum_{m_n \in \{0,1\}} \alpha^{m_n} A(t_{n-1}) A(t_n)^{m_n} \\
&\quad + e^{-b(t_n - t_{n-2})} \sum_{m_{n-1}, m_n \in \{0,1\}} \alpha^{m_{n-1} + m_n} A(t_{n-2}) A(t_{n-1})^{m_{n-1}} A(t_n)^{m_n} \\
&\quad + \cdots \\
&\quad + e^{-b(t_n - t_1)} \sum_{m_2, \dots, m_n \in \{0,1\}} \alpha^{\sum_{i=2}^{n} m_i} A(t_1) A(t_2)^{m_2} \cdots A(t_n)^{m_n}. \quad (4.24)
\end{aligned}
$$

The win score that v_j has gained against v_i is contained in $(W(t_n))_{ij}$. The dynamic win score is represented in the vector form as

$$
\boldsymbol{w}(t_n) = \mathbf{1} W(t_n), \quad (4.25)
$$

where $\boldsymbol{w}(t_n)$ is a row vector whose ith element is equal to the win score of v_i at time t_n. In Eq. (4.24), the first term on the right-hand side represents the effect of direct wins. The remaining terms are not classified according to the length of the indirect wins, as is the case in Eq. (4.22), but to the time when a chain of wins starts. The second term is contributed by direct wins at t_{n-1} (when $m_n = 0$) and indirect wins of length two, one at t_{n-1} and the other at t_n (when $m_n = 1$). The direct and indirect wins in the past are discounted exponentially in time such that the first win in a chain of wins specifies the time of the chain. Because the chain starts at t_{n-1} and the current time is t_n, we have a factor $e^{-b(t_n - t_{n-1})}$ in the second term on the right-hand side of Eq. (4.24). Similarly, four types of chains enumerated in the third term, identified by $(m_{n-1}, m_n) = (0, 0)$, $(0, 1)$, $(1, 0)$, and $(1, 1)$, start at t_{n-2} and are subjected to the discount factor $e^{-b(t_n - t_{n-2})}$.

Because Eq. (4.24) is rewritten as

$$W(t_n)$$

$$= A(t_n) + e^{-b(t_n - t_{n-1})} \times \sum_{m_n \in \{0,1\}} \alpha^{m_n} A(t_n)^{m_n} \times$$

$$\left[A(t_{n-1}) \right.$$

$$+ e^{-b(t_{n-1} - t_{n-2})} \sum_{m_{n-1} \in \{0,1\}} \alpha^{m_{n-1}} A(t_{n-2}) A(t_{n-1})^{m_{n-1}}$$

$$+ \cdots$$

$$\left. + e^{-b(t_{n-1} - t_1)} \sum_{m_2,\dots,m_{n-1} \in \{0,1\}} \alpha^{\sum_{i=2}^{n-1} m_i} A(t_1) A(t_2)^{m_2} \cdots A(t_{n-1})^{m_{n-1}} \right]$$

$$= A(t_n) + e^{-b(t_n - t_{n-1})} W(t_{n-1}) \left[I + \alpha A(t_n) \right], \tag{4.26}$$

we obtain the online update equation for the dynamic win-lose score:

$$\boldsymbol{w}(t_n) = \begin{cases} 1 A(t_1) & (n = 1), \\ 1 A(t_n) + e^{-b(t_n - t_{n-1})} \boldsymbol{w}_{n-1}(I + \alpha A(t_n)) & (n > 1). \end{cases} \tag{4.27}$$

The definition of the dynamic lose score is similar to that of the dynamic win score. The dynamic win-lose score is equal to the difference between the dynamic win score and the dynamic lose score.

The dynamic win-lose score has been used to rank professional men's tennis players across time [Motegi and Masuda (2012)]. It yields a time-varying strength of each player, similar to time courses of the ATP ranking, which is the official ranking. The dynamic win-lose score also predicts the result of the match with a higher accuracy than the static methods and the official ranking do, where a prediction is regarded to be correct if a higher ranker defeats a lower ranker.

4.6 Statistical properties of event times

Consider a temporal network in the event-based representation. Because we have discarded the duration of each event for simplicity, the event sequence associated with a given node or link is given by $\{t_1, t_2, \dots, t_n\}$. Differently

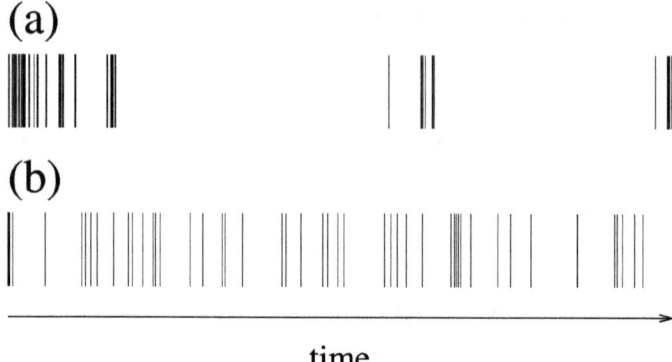

time

Fig. 4.10   (a) Bursty and (b) non-bursty event sequences. The two event sequences have the same number of events, $n = 50$.

from Eq. (4.2), here we interpret t_i and n as being quantities for a single node or link, not for the entire network.

In this section, we introduce several measurements for such time-stamped event sequences. Analysis of distributions of inter-event times suggests that Poisson processes are far from most empirical data and we should use, for example, renewal processes (Section 4.6.1). The coefficient of variation (Section 4.6.2) and Fano factor (Section 4.6.5) implicitly assume that an event sequence is stationary, or more practically, a renewal process. However, empirical event sequences often deviate from renewal processes and are non-stationary. When an event sequence is non-stationary, we should additionally use other methods, such as the local variation (Section 4.6.3), detrending (Section 4.6.4) and detrended fluctuation analysis (Section 4.6.6). We introduce measurements for temporal correlation in Section 4.7.

4.6.1   *Distribution of inter-event times*

Event sequences generated from a Poisson process do not approximate $\{t_1, t_2, \ldots, t_n\}$ observed in various empirical data. In particular, real event sequences are often bursty (Fig. 4.10(a)). Many events occur within a short interval, generating a burst, while different bursts tend to be separated by a long interval. This bursty feature is lacking in event sequences generated from a Poisson process with the same average rate (Fig. 4.10(b)).

To be quantitative, we define the inter-event time as the time between two consecutive events on the link. The ith inter-event time, denoted by τ_i

$(1 \leq i \leq n - 1)$, is defined by

$$\tau_i = t_{i+1} - t_i. \tag{4.28}$$

The probability density function of inter-event times, τ, for a Poisson process is given by the exponential distribution

$$\psi(\tau) = \lambda e^{-\lambda \tau}, \tag{4.29}$$

where λ is the parameter (Eq. (2.31)). In contrast, many real data yield a power-law distribution of inter-event times, i.e.,

$$\psi(\tau) \propto \tau^{-\alpha} \tag{4.30}$$

for $\alpha > 0$, at least up to some cut-off value of τ. Empirical observations suggest that $1 \leq \alpha \leq 2.7$ in various data [Vázquez *et al.* (2006); Karsai *et al.* (2012)].

4.6.2 *Coefficient of variation*

The bursty event sequence shown in Fig. 4.10(a) shows a large variance of the inter-event time as compared to the Poisson processes (Fig. 4.10(b)). The coefficient of variation (CV) for the inter-event time is defined as the standard deviation of $\{\tau_i\}$ divided by its mean, i.e.,

$$CV = \frac{\sqrt{\frac{1}{n-1} \sum_{i=1}^{n-1} (\tau_i - \langle \tau \rangle)^2}}{\langle \tau \rangle}, \tag{4.31}$$

where

$$\langle \tau \rangle = \frac{1}{n-1} \sum_{i=1}^{n-1} \tau_i \tag{4.32}$$

is the average inter-event time. It should be noted that the number of inter-event times is equal to $n - 1$.

The exponential distribution, corresponding to a Poisson process, produces $CV = 1$ because both mean and standard deviation are equal to $1/\lambda$. A periodic event sequence yields $CV = 0$ because τ_i is the same for all i. A long-tailed distribution yields a large value of CV. The CV for the inter-event time has been used for neuronal spike trains since early 1990s [Softky and Koch (1993)]. For neurons, CV values are often somewhat larger than unity, but not as large as values for typical power-law distributions.

A normalised variant of the CV, called the burstiness, is often used in network science community [Goh and Barabási (2008)]. The burstiness, denoted by B, is defined by

$$B = \frac{CV - 1}{CV + 1} \tag{4.33}$$

and verifies $-1 \leq B < 1$. Poisson processes (i.e., CV = 1) yield $B = 0$. A periodic event sequence (i.e., CV = 0) yields $B = -1$. An extremely bursty event sequence, i.e., $CV \to \infty$, yields $B \to 1$. Many data sets yield substantially positive values of B [Goh and Barabási (2008)].

4.6.3 *Local variation*

The CV is considerably affected by the presence of a small fraction of extremely large inter-event times and by non-stationarity of the time series. A similar measure that does not suffer from this problem and is potentially useful for the analysis of temporal network data is the local variation (LV), originally introduced in the analysis of neuronal spike trains [Shinomoto *et al.* (2003)]. The LV is defined by

$$\text{LV} = \frac{3}{n-1} \sum_{i=1}^{n-1} \left(\frac{\tau_i - \tau_{i+1}}{\tau_i + \tau_{i+1}} \right)^2. \tag{4.34}$$

With this normalisation factor $3/(n-1)$, a Poisson process and periodic event sequence yield LV = 1 and LV = 0, respectively. A bursty event sequence yields a large LV value. It holds true that $0 \leq \text{LV} < 3$.

Even if an inter-event time is extremely large, its effect is confined in two consecutive terms on the right-hand side of Eq. (4.34), therefore the name "local". Note that each term on the right-hand side is at most unity, whereas a single term in the CV can be indefinitely large. For the non-homogeneous Poisson process with rate function $\lambda(t)$, the effect of the fluctuating event rate on variation in the inter-event time is eliminated by the divisive factor $(\tau_i + \tau_{i+1})^2$ in the definition of the LV. This fact is more visible if we rewrite

$$\left(\frac{\tau_i - \tau_{i+1}}{\tau_i + \tau_{i+1}} \right)^2 = 1 - \frac{4\tau_i \tau_{i+1}}{\left(\tau_i + \tau_{i+1} \right)^2}. \tag{4.35}$$

A non-homogeneous Poisson process would yield a LV value close to unity while its value of CV can be substantially larger than unity. It should be noted that, unlike the CV, the LV is affected by temporal correlation between inter-event times, which will be discussed in Section 4.7.

4.6.4 *Detrending*

Human-related activities are usually affected by circadian and weekly rhythms. A large fluctuation in the inter-event time, which signifies bursti-ness, may be simply due to strong periodic rhythms. One method with which to treat this problem is to use non-homogeneous Poisson processes whose rates are periodically modulated daily and weekly [Malmgren *et al.* (2008)]. Another method, which we explain here, is to detrend event sequences to get rid of rhythmic effects and then investigate burstiness by, for example, measuring the CV [Anteneodo *et al.* (2010); Jo *et al.* (2012)].

Denote by $n(t)$ the number of events at discretised time t $(1 \leq t \leq t_{\max})$, akin to the snapshot representation. Given the period of the cycle t_{p} in terms of the discretised time (e.g., one day or one week), the event rate $\lambda(t)$ $(1 \leq t \leq t_{\mathrm{p}})$ within the period is defined by

$$\lambda(t) = \frac{t_{\mathrm{p}}}{\sum_{t'=1}^{t_{\max}} n(t')} \sum_{k=0}^{\lfloor t_{\max}/t_{\mathrm{p}} \rfloor} n(t + kt_{\mathrm{p}}), \qquad (4.36)$$

where $\sum_{t'=1}^{t_{\max}} n(t')$ is the total number of events. The normalisation constant on the right-hand side in Eq. (4.36) respects

$$\frac{\sum_{t=1}^{t_{\mathrm{p}}} \lambda(t)}{t_{\mathrm{p}}} \approx 1. \qquad (4.37)$$

In other words, the mean event rate within the period is approximately equal to unity. It is exactly equal to unity if $t_{\max}$ is a multiple of t_{p}.

We define the rescaled time $t^*(t)$ by

$$t^*(t) = \sum_{1 \leq t' \leq t} \lambda(t') \quad (0 \leq t \leq t_{\mathrm{p}}). \qquad (4.38)$$

Because $t^*(t) \approx \int_0^t \mathrm{d}t' \lambda(t')$, Eq. (4.38) corresponds to the change of the time variable by $\lambda^*(t^*)\mathrm{d}t^* = \lambda(t)\mathrm{d}t$, where $\lambda^*(t^*) = 1$. Therefore, the rhythm with period t_{p} has been detrended if we use the rescaled time t^* to define the inter-event time. It should be noted that the combination of Eqs. (4.37) and (4.38) yields $t^*(t_{\mathrm{p}}) = t_{\mathrm{p}}$. Therefore, the rescaled time coincides with the original time at $t^* = t = t_{\mathrm{p}}$, which is identified with $t^* = t = 0$.

The rescaled inter-event time is defined by

$$\tau_i^* = t^*(t_{i+1}) - t^*(t_i) = \sum_{t_i < t' \leq t_{i+1}} \lambda(t'), \qquad (4.39)$$

where, to circumvent the periodic boundary condition, we add t_{p} to $t^*(t_{i+1})$ if $t^*(t_i)$ is larger than $t^*(t_{i+1})$. For a mobile phone data set, the burstiness persists even under the rescaled time [Jo *et al.* (2012)], indicating that the burstiness observed in the original data is not entirely due to trivial periodicity of human activity.

4.6.5   *Fano factor*

For a given time window $[0, t_{max}]$, the Fano factor is defined by

$$\text{Fano}(t_{max}) = \frac{(\text{variance of the number of events in } [0, t_{max}])}{(\text{mean number of events in } [0, t_{max}])}. \quad (4.40)$$

The Fano factor is equal to unity for Poisson processes for any rate λ and any t_{max} because the number of events in $[0, t_{max}]$ obeys the Poisson distribution with mean and variance both equal to λt_{max} (Section 2.2.1). Non-homogeneous Poisson processes also yield $\text{Fano}(t_{max}) = 1$. If the process is deterministic, including periodic sequences, we obtain $\text{Fano}(t_{max}) = 0$.

The Fano factor for the renewal process with a power-law distribution of inter-event times $\psi(\tau) \propto \tau^{-\mu}$ is derived as follows. We recall that $n(t)$ represents the number of events in $[0, t]$. We assume that an event occurs at time 0, which defines the so-called ordinary renewal process (see Section 6.1 for more about ordinary renewal processes).

By substituting Eqs. (2.76), the Laplace transform of the power-law distribution, in (2.48), we obtain

$$\langle \hat{n}(s) \rangle = \begin{cases} -\frac{1}{\Gamma(1-\mu)} s^{-\mu} + o(s^{-\mu}) & (1 < \mu < 2), \\ \frac{1}{\langle t \rangle} s^{-2} + \frac{\Gamma(1-\mu)}{\langle t \rangle^2} s^{\mu-4} + o(s^{\mu-4}) & (2 < \mu < 3), \\ \frac{1}{\langle t \rangle} s^{-2} + \frac{\langle t^2 \rangle - 2\langle t \rangle^2}{2\langle t \rangle^2} s^{-1} + \frac{\Gamma(1-\mu)}{\langle t \rangle^2} s^{\mu-4} + o(s^{\mu-4}) & (3 < \mu < 4). \end{cases}$$
$$(4.41)$$

By applying the Tauberian theorem to Eq. (4.41), we obtain [Feller (1949); Smith (1961); Teugels (1968)]

$$\langle n(t) \rangle = \begin{cases} -\frac{1}{\Gamma(1-\mu)\Gamma(\mu)} t^{\mu-1} + o(t^{\mu-1}) & (1 < \mu < 2), \\ \frac{t}{\langle t \rangle} + \frac{\Gamma(1-\mu)}{\Gamma(4-\mu)\langle t \rangle^2} t^{-\mu+3} + o(t^{-\mu+3}) & (2 < \mu < 3), \\ \frac{t}{\langle t \rangle} + \frac{\langle t^2 \rangle - 2\langle t \rangle^2}{2\langle t \rangle^2} + \frac{\Gamma(1-\mu)}{\Gamma(4-\mu)\langle t \rangle^2} t^{-\mu+3} + o(t^{-\mu+3}) & (3 < \mu < 4). \end{cases}$$
$$(4.42)$$

For $\mu \geq 4$, the first two leading terms stay the same as those for $3 < \mu < 4$.

To validate the theoretical results, Eq. (4.42), we numerically simulate a renewal process whose distribution of inter-event times is given by

$$\psi(\tau) = \frac{\mu - 1}{(t + 1)^{\mu}}, \quad (4.43)$$

which has the mean $\langle \tau \rangle = 1/(\mu - 2)$ when $\mu > 2$. The power-law coefficient α' for the number of events as a function of time, i.e., $\langle n(t) \rangle \propto t^{\alpha'}$, is shown in Fig. 4.11(a). The figure indicates that Eq. (4.42) explains the numerical results reasonably well, particularly when μ is not close to two.

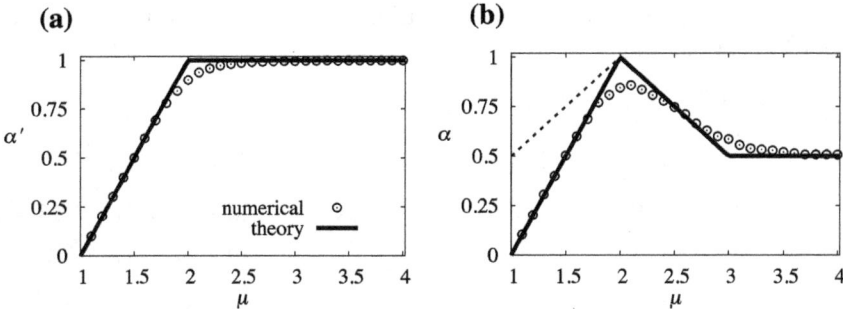

Fig. 4.11 Scaling of the number of events in the renewal process whose inter-event time is distributed according to $\psi(t) \propto t^{-\mu}$. (a) α' defined via $\langle n(t) \rangle \propto t^{\alpha'}$ as a function of μ. (b) α defined via $\sigma(n(t)) \propto t^{\alpha}$ as a function of μ. The numerical results are shown by the circles. The solid lines in (a) and (b) represent the theory obtained by Eqs. (4.42) and (4.47), respectively. The dotted line in (b) represents the theoretical prediction of the detrended fluctuation analysis, $F(\ell) \propto \ell^{\alpha}$ [Allegrini *et al.* (2009); Buldyrev (2010); Rybski *et al.* (2012)]. For each value of μ, we generated 5×10^3 event sequences and calculated $\langle n(t) \rangle$ and $\sigma(n(t))$ for 15 values of t that were equally distant in the log space. Each event sequence lasted until $t = 3.17 \times 10^5$ at $\mu = 3$. For $\mu > 2$, we rescaled the time to work with $t' \to (\mu - 2)t$ to keep the effective $\langle t \rangle$ the same for different values of μ. For $\mu \leq 2$, for which $\langle t \rangle$ diverges, we manually rescaled the time to ensure that $\langle n(t) \rangle > 10^3$ for $\mu \geq 1.3$ and $\langle n(t) \rangle > 10^2$ for $\mu = 1.2$ and $\langle n(t) \rangle > 20$ for $\mu = 1.1$. Then, for each value of μ, we fitted a line by the least square method in the t-$\langle n(t) \rangle$ or t-$\sigma(n(t))$ plane on the logarithmic scale. In all cases, the Pearson's correlation coefficient between $\log \langle n(t) \rangle$ and $\log t$ and that between $\log \sigma(n(t))$ and $\log t$ were larger than 0.999.

To calculate the second moment, which comes to the numerator of the Fano factor, we substitute Eq. (2.76) in Eq. (2.49) to obtain

$$
\langle \hat{n^2}(s) \rangle = \begin{cases} \frac{2}{\Gamma(1-\mu)^2} s^{-2\mu+1} + o(s^{-2\mu+1}) & (1 < \mu < 2), \\ \frac{2}{\langle t \rangle^2} s^{-3} - \frac{4\Gamma(1-\mu)}{\langle t \rangle^3} s^{\mu-5} + o(s^{\mu-5}) & (2 < \mu < 3), \\ \frac{2}{\langle t \rangle^2} s^{-3} + \frac{2\langle t^2 \rangle - 3\langle t \rangle^2}{\langle t \rangle^3} s^{-2} + \frac{4\Gamma(1-\mu)}{\langle t \rangle^3} s^{\mu-5} + o(s^{\mu-5}) & (3 < \mu < 4). \end{cases}
$$

$$(4.44)$$

By applying the Tauberian theorem, we obtain [Feller (1949); Teugels (1968)]

$$
\langle n^2(t) \rangle = \begin{cases} \frac{2}{\Gamma(1-\mu)^2 \Gamma(2\mu-1)} t^{2\mu-2} + o(t^{2\mu-2}) & (1 < \mu < 2), \\ \frac{t^2}{\langle t \rangle^2} - \frac{4\Gamma(1-\mu)}{\Gamma(5-\mu)\langle t \rangle^3} t^{-\mu+4} + o(t^{-\mu+4}) & (2 < \mu < 3), \\ \frac{t^2}{\langle t \rangle^2} + \frac{2\langle t^2 \rangle - 3\langle t \rangle^2}{\langle t \rangle^3} t + \frac{4\Gamma(1-\mu)}{\Gamma(4-\mu)\langle t \rangle^3} t^{4-\mu} + o(t^{4-\mu}) & (3 < \mu < 4). \end{cases}
$$

$$(4.45)$$

Using Eqs. (4.42) and (4.45), the variance of $n(t)$ is given to the leading order in t as follows:

$$[\sigma(n(t))]^2 = \langle n^2(t) \rangle - \langle n(t) \rangle^2$$

$$= \begin{cases} -\frac{1}{\Gamma(1-\mu)} \left[\frac{2}{\Gamma(2\mu-1)} + \frac{1}{\Gamma(\mu)^2} \right] t^{2\mu-2} & (1 < \mu < 2), \\ \frac{2\Gamma(1-\mu)}{\langle t \rangle^3} \left[\frac{2}{\Gamma(5-\mu)} - \frac{1}{\Gamma(4-\mu)} \right] t^{-\mu+4} & (2 < \mu < 3), \\ \frac{\langle t^2 \rangle - \langle t \rangle^2}{\langle t \rangle^3} t & (3 < \mu < 4), \end{cases} \quad (4.46)$$

where $\sigma(n(t))$ denotes the standard deviation of $n(t)$. The results for $3 < \mu < 4$ coincide with the results for general $\psi(t)$ with a finite variance derived by the Gaussian approximation [Cox (1962)]. This result holds true for larger values of μ because the leading terms for $\mu \geq 4$ remain the same as those for $3 < \mu < 4$.

Therefore, we conclude $\sigma(n(t)) \propto t^\alpha$, where

$$\alpha = \begin{cases} \mu - 1 & (1 < \mu < 2), \\ 2 - \mu/2 & (2 < \mu < 3), \\ 1/2 & (\mu > 3). \end{cases} \quad (4.47)$$

The numerically obtained power-law coefficient α for various values of μ is shown by the circles in Fig. 4.11(b), reasonably agreeing with the theoretical estimate by Eq. (4.47) shown by the solid line.

On the basis of Eqs. (4.42) and (4.46), the leading term of the Fano factor in terms of t is given by

$$\text{Fano}(t) \propto \begin{cases} t^{\mu-1} + o(t^{\mu-1}) & (1 < \mu < 2), \\ t^{-\mu+3} + o(t^{-\mu+3}) & (2 < \mu < 3), \\ \frac{\langle t^2 \rangle - \langle t \rangle^2}{\langle t \rangle^2} + o(1) & (\mu > 3). \end{cases} \quad (4.48)$$

The result for $\mu > 3$ is consistent with that for Poisson processes, i.e., $\text{Fano}(t) = 1$, with $\langle t^2 \rangle - \langle t \rangle^2 = 1/\lambda^2$ and $\langle t \rangle = 1/\lambda$. Renewal processes with $\psi(\tau) \propto \tau^{-\mu}$ ($\mu > 3$) are qualitatively the same as Poisson processes as far as the Fano factor is concerned.

To calculate the Fano factor for given data, we should consider an ensemble of event sequences by, for example, using sequences from different individuals or dividing a single sequence into some sequences and regard them as an ensemble. The same treatment is done in the detrended fluctuation analysis explained in the next section.

4.6.6 *Detrended fluctuation analysis*

Detrended fluctuation analysis (DFA) is a scaling analysis method for time series data to investigate long-range temporal correlation. DFA has been applied to various types of time series data including DNA, heart dynamics and weather records, just to name a few [Peng *et al.* (1994); Hu *et al.* (2001); Kantelhardt *et al.* (2001)]. Differently from the Fano factor, which is also a fluctuation analysis method, DFA detrends the time series before the fluctuation analysis is carried out, as the name stands. DFA is based on random walk theory and works for general time series data, not only for event sequence data.

Consider a time series equidistantly measured in discrete time, $\{x(t); t = 1, 2, \ldots, t_{\max}\}$. The autocorrelation function for $\{x(t)\}$ is given by

$$C(\Delta t) = \frac{1}{t_{\max} - \Delta t} \sum_{t=1}^{t_{\max} - \Delta t} [x(t) - \langle x \rangle] [x(t + \Delta t) - \langle x \rangle], \qquad (4.49)$$

where $\langle x \rangle = \sum_{t=1}^{t_{\max}} x(t)/t_{\max}$. We say that $\{x(t)\}$ has short-range correlation if $C(\Delta t) \propto \exp(-\Delta t/t_0)$ for some $t_0 > 0$ and long-range correlation if $C(\Delta t) \propto (\Delta t)^{-\gamma}$, where $0 < \gamma < 1$. Calculating $C(\Delta t)$ directly from Eq. (4.49) is often difficult due to the trend and other non-stationarity that may be present in the data.

DFA, a bit simplified for the purpose of the demonstration here, consists of the following steps.

(1) The trajectory of the random walk whose increment is determined by $\{x(t)\}$ is given by

$$Y(t) = \sum_{t'=1}^{t} x(t'). \qquad (4.50)$$

(2) We divide time series $\{Y(t)\}$ into non-overlapping segments of equal length ℓ.
(3) Determine the local trend for each segment by a least-squares fitting of a polynomial to the segment. If the fitted polynomial is of the nth order, we refer to the method as DFAn. DFA2 is a common choice.
(4) Detrend the time series by subtracting the local trend from $\{Y(t)\}$ to obtain the detrended time series $\{Y_\ell(t)\}$. For example, $Y_\ell(t) = Y(t) - f_1(t)$ $(1 \leq t \leq \ell)$, where $\{f_1(t)\}$ $(1 \leq t \leq \ell)$ is the local trend for the first segment, i.e., $\{Y(t); 1 \leq t \leq \ell\}$. Similarly, $Y_\ell(t) = Y(t) - f_2(t)$ $(\ell + 1 \leq t \leq 2\ell)$, where $\{f_2(t)\}$ $(\ell + 1 \leq t \leq 2\ell)$ is the local trend for the second segment.

(5) The DFA fluctuation function is given as the square root of the average variance of the detrended time series, i.e.,

$$F(\ell) = \sqrt{\frac{1}{t_{\max}} \sum_{t=1}^{t_{\max}} [Y_\ell(t)]^2}. \tag{4.51}$$

The standard deviation $F(\ell)$ increases as the duration of the segment, ℓ, increases. If $\{x(t)\}$ has long-range correlation, $F(\ell)$ increases by a power law as $F(\ell) \propto \ell^\alpha$ such that the power-law exponent α is of interest.

For DFA1, let us illustrate the calculation of $F(\ell)$ [Taqqu *et al.* (1995); Kantelhardt *et al.* (2001)]. We need to calculate

$$[F(\ell)]^2 = \langle \frac{1}{\ell} \sum_{t=1}^{\ell} [Y_\ell(t)]^2 \rangle = \frac{1}{\ell} \sum_{t=1}^{\ell} \langle [Y(t) - c_1 t - c_2]^2 \rangle, \tag{4.52}$$

where $f_1(t) = c_1 t + c_2$ is the local trend. Coefficients c_1 and c_2 are obtained from the least squares fitting and hence represented as linear sums of $Y(t)$ ($1 \le t \le \ell$). Therefore, what we need to calculate Eq. (4.52) is the mean and covariance of $Y(t)$, i.e., $\langle Y(t) \rangle$ and $\langle Y(t_1)Y(t_2) \rangle$, where $1 \le t, t_1, t_2 \le \ell$. In fact, the former will be unneeded [Taqqu *et al.* (1995)]. By assuming $t_1 \le t_2$ without loss of generality, we obtain the covariance as follows:

$$\langle Y(t_1)Y(t_2) \rangle$$

$$= \langle \sum_{t_1'=1}^{t_1} x(t_1') \sum_{t_2'=1}^{t_2} x(t_2') \rangle$$

$$= C(-t_1 + 1) + 2C(-t_1 + 2) + \cdots + (t_1 - 1)C(-1)$$
$$+ t_1 C(0) + t_1 C(1) + \cdots + t_1 C(t_2 - t_1)$$
$$+ (t_1 - 1)C(t_2 - t_1 + 1) + (t_1 - 2)C(t_2 - t_1 + 2) + \cdots + C(t_2 - 1)$$

$$= \sum_{\Delta t=1}^{t_1-1} (t_1 - \Delta t)C(\Delta t) + t_1 \sum_{\Delta t=0}^{t_2-t_1} C(\Delta t) + \sum_{\Delta t=t_2-t_1+1}^{t_2-1} (t_2 - \Delta t)C(\Delta t). \tag{4.53}$$

When $\{x(t)\}$ has long-range correlation, we use the approximation

$$\sum_{\Delta t=1}^{t-1} C(\Delta t) \propto \sum_{\Delta t=1}^{t} (\Delta t)^{-\gamma} \approx \int_1^t (\Delta t)^{-\gamma} \, d\Delta t \propto t^{1-\gamma} \tag{4.54}$$

for large Δt. Similarly, we obtain

$$\sum_{\Delta t=1}^{t-1} \Delta t C(\Delta t) \propto t^{2-\gamma}. \tag{4.55}$$

By substituting Eqs. (4.53), (4.54) and (4.55) in Eq. (4.52), after some calculus, we obtain

$$F(\ell) \propto \ell^{1-\gamma/2} \quad (0 < \gamma < 1). \tag{4.56}$$

When $\gamma > 1$ or $C(t)$ decays exponentially, i.e., the case of short-range correlation, we obtain $F(\ell) \propto \ell^{1/2}$ [Peng *et al.* (1994)].

DFA has been applied to sequences of events whose interval obeys a long-tailed distribution in a uncorrelated or correlated manner. Rybski and colleagues theoretically derived the power-law exponent α for $F(\ell) \propto \ell^{\alpha}$ when inter-event times are independently distributed according to the power law, $\psi(\tau) \propto \tau^{-\mu}$ [Buldyrev (2010); Rybski *et al.* (2012)] (also see [Allegrini *et al.* (2009)]). The α values that they derived are the same as that for the Fano factor except for the case $1 < \mu < 2$. The α value for DFA when $1 < \mu < 2$ is plotted by the dotted line in Fig. 4.11(b). These results indicate that DFA can yield a coefficient corresponding to long-range correlation (i.e., $1/2 < \alpha < 1$) even if there is no correlation in inter-event times. Rybski and colleagues analysed message sending activity of users in an online community [Rybski *et al.* (2012)]. They found that shuffling the inter-event times did not change α for data for individual users. This result indicates that the anomalous α values (i.e., $1/2 < \alpha < 1$) obtained from DFA are caused by long-tailed behaviour of τ. In contrast, they also found that shuffling the inter-event times changed α when DFA was applied to the event sequence obtained as the aggregation over different users. This result indicates that the anomalous α values observed in this case are at least partly due to temporal correlation present in the data.

4.7 Temporal correlation

In real temporal networks, networks observed at different times are usually correlated in various ways. For example, if link (v_i, v_j) has an event at time t after a short inter-event time, the same link tends to have a next event after a short inter-event time after t. In addition, from v_i's point of view, a short inter-event time may induce a next event with some other node $v_{j'} (\neq j)$ within a short time. In contrast, if the last inter-event time has been long, the inter-event time to the next event may be long, too. Such positive correlation in inter-event time is prevalent in empirical data, even after removing the effect of circadian and other rhythms [Karsai *et al.* (2012)]. It should be noted that correlation in inter-event time contradicts the premise of the renewal process, in which inter-event times are independent by definition. Some generative models of temporal networks take temporal correlation into account by assuming that a snapshot in discrete time $t + 1$ depends on the snapshot at time t (e.g., Section 5.6). In this section, we explain measures of correlation in temporal networks.

Let us start with the case of a single event sequence, associated with a single node or link. A succinct way to visualise possible correlation is to plot the conditional mean inter-event time defined by

$$\tau^{\text{next}}(\tau) \equiv \langle \tau_{i+1} \rangle_{\tau_i \leq \tau}. \tag{4.57}$$

The mean on the right-hand side is conditioned by $\tau_i \leq \tau$ to make a smooth plot. If different inter-event times are uncorrelated, $\tau^{\text{next}}(\tau)$ is independent of τ. In various data sets, $\tau^{\text{next}}(\tau)$ increases as a function of τ, implying positive correlation [Masuda *et al.* (2013b)].

A popularly used measure of the correlation is the memory coefficient [Goh and Barabási (2008)], often denoted by M, defined by

$$\frac{\sum_{i=1}^{n-2}(\tau_i - m_1)(\tau_{i+1} - m_2)}{\sqrt{\sum_{i=1}^{n-2}(\tau_i - m_1)^2 \sum_{i=2}^{n-1}(\tau_i - m_2)^2}}, \tag{4.58}$$

where

$$m_1 = \frac{1}{n-2} \sum_{i=1}^{n-2} \tau_i, \tag{4.59}$$

$$m_2 = \frac{1}{n-2} \sum_{i=2}^{n-1} \tau_i. \tag{4.60}$$

The memory coefficient is the sample covariance of adjacent inter-event times. The Pearson correlation coefficient for $\{\tau_i\}$ is equal to the square root of the memory coefficient. Many empirical data sets possess a positive memory coefficient, implying positive correlation in inter-event times [Goh and Barabási (2008)].

The memory coefficient can be considerably affected by the presence of extremely long inter-event times, which is typical in empirical data. Therefore, it may be useful to turn to Spearman/Kendall rank coefficients.

Let us turn to a different measure of correlation that is insensitive to extremely long inter-event times [Karsai *et al.* (2012)]. We define a burst by a sequence of events such that the inter-event times between consecutive events within the sequence are at most $\Delta\tau$ and those between the first event in the sequence and the preceding event and between the last event

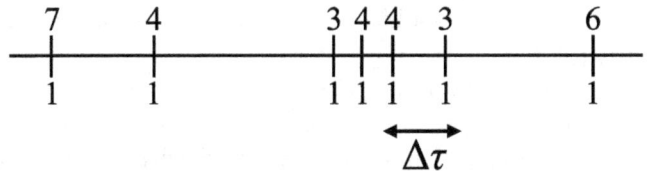

Fig. 4.12   An event sequence. The number indicates the node index.

in the sequence and the following event are larger than $\Delta\tau$. For example, in Fig. 4.12, the length of the burst containing the third event is equal to four. If the inter-event time is independent of each other, the probability that a burst contains n_b (≥ 0) events obeys the geometric distribution given by

$$p(n_b) = \left[\int_0^{\Delta\tau} \psi(\tau)d\tau \right]^{n_b-1} \left[1 - \int_0^{\Delta\tau} \psi(\tau)d\tau \right]. \qquad (4.61)$$

This is because $\int_0^{\Delta\tau} \psi(\tau)d\tau$ represents the probability that the next event occurs within time $\Delta\tau$ since the previous event. It should be noted that this expression is not affected by whether the distribution of inter-event times, $\psi(\tau)$, is long- or short-tailed.

Equation (4.61) implies that $p(n_b)$ has a short tail. However, for various real data and a wide range of $\Delta\tau$, $p(n_b)$ obeys a power-law distribution $p(n_b) \propto n_b^{-\beta}$, where β is somewhere between 2.3 and 4.1 depending on data [Karsai *et al.* (2012)]. Therefore, bursts containing many events are more likely to appear in real data than hypothetical independent sequences of inter-event times. This result implies that a short inter-event time tends to succeed after a short one to make the burst size large. We emphasise that this result is independent of the shape of $\psi(\tau)$. Logically, $p(n_b)$ can be a power law even for an event sequence whose inter-event time obeys a short-tailed distribution.

There are similar correlation measures for the snapshot representation of temporal networks [Clauset and Eagle (2007); Tang *et al.* (2010b)]. For example, the temporal correlation coefficient [Tang *et al.* (2010b)] for two adjacent snapshots is defined by

$$C = \frac{1}{N(t_{\max}-1)} \sum_{t=1}^{t_{\max}-1} \sum_{i=1}^{N} \frac{\sum_{j=1}^{N} A_{ij}(t)A_{ij}(t+1)}{\sqrt{\left[\sum_{j=1}^{N} A_{ij}(t)\right]\left[\sum_{j=1}^{N} A_{ij}(t+1)\right]}}, \qquad (4.62)$$

where we remind that $t_{\max}$ the number of snapshots. C ranges between 0 and 1, and $C = 1$ if and only if all snapshots are the same (i.e., static network). The value of C would be large if many links are used at both t and $t+1$ $(1 \leq t \leq t_{\max} - 1)$.

The measures of correlation introduced so far focus on a single node or link. Therefore, they are ignorant of correlation across different nodes or links. For example, they are not concerned with the possibility that an event on link (v_i, v_j) increases the likelihood of an event on link $(v_i, v_{j'})$ $(j' \neq j)$ shortly after. In the rest of this section, we introduce measurements for such correlation.

We can measure temporal correlation across links by comparing the unconditional probability distribution of neighbouring nodes with that conditioned by the previous neighbour. Here we use entropy and mutual information to do this [Takaguchi *et al.* (2011)], hinted by analysis of human mobility patterns [Song *et al.* (2010)].

Consider an event sequence for the ith node. We neglect inter-event times and assume for simplicity that the ith node is adjacent to one node after another. For example, Fig. 4.12 illustrates the events that node v_1 experiences. We say that the partner sequence for this node is $\{v_7, v_4, v_3, v_4, v_4, v_3, v_6\}$. If the node is simultaneously adjacent to multiple nodes at the same time, which may happen in the snapshot representation of temporal networks, we break ties by randomly ordering nodes that are simultaneously adjacent to the ith node. The entropy associated with the distribution of adjacent nodes of the ith node is given by

$$H_i^1 = - \sum_{j=1; j\neq i}^{N} p_i(j) \log_2 p_i(j), \qquad (4.63)$$

where $p_i(j)$ is the probability that the ith node is adjacent to the jth node. If the ith node has a large degree in the aggregate network and is equally connected to different nodes, H_i^1 is large.

The entropy conditioned by the previously adjacent node is defined by

$$H_i^2 = - \sum_{j=1; j\neq i}^{N} p_i(j) \sum_{\ell=1; \ell\neq i}^{N} p_i(\ell|j) \log_2 p_i(\ell|j), \qquad (4.64)$$

where $p_i(\ell|j)$ is the probability that the ith node is adjacent to the ℓth node under the condition that the ith node was adjacent to the jth node in the previous snapshot. H_i^2 measures the second-order correlation in the sequence of neighbours of the ith node and represents the uncertainty about the next neighbour that remains in the presence of the knowledge about the previous neighbour. For any i, $0 \leq H_i^2 \leq H_i^1 \leq \log_2 N$ holds true.

The mutual information is given by

$$I_i \equiv H_i^1 - H_i^2 = \sum_{j,\ell=1;j,\ell \neq i}^{N} p_i(j,\ell) \log_2 \frac{p_i(j,\ell)}{p_i(\ell)p_i(j)}, \qquad (4.65)$$

where $p_i(j,\ell)$ represents the joint probability that the ith node is adjacent to the jth node and then the ℓth node immediately after. Mutual information I_i represents the information about the next neighbour gained by knowing the present neighbour.

The absence of temporal correlation implies $I_i = 0$ because knowing the present neighbour would not give any clue about the next neighbour. In contrast, a positive value of I_i implies the presence of temporal correlation. The mutual information of most nodes is significantly positive for some empirical temporal networks. Burstiness is a dominant reason to make I_i positive. However, even after we control for the effect of burstiness, I_i remains significantly positive [Takaguchi *et al.* (2011)].

Temporal correlation across different links sharing a node at a network level can be measured by the betweenness preference [Pfitzner *et al.* (2013)]. Consider a possibly directed temporal network in the snapshot representation and define

$$B_{j\ell}^i(t) = \begin{cases} 1 & (A_{ji}(t-1) = A_{i\ell}(t) = 1), \\ 0 & (\text{otherwise}). \end{cases} \qquad (4.66)$$

$B_{j\ell}^i(t)$ is equal to unity if v_i is between v_j and v_ℓ on a temporal walk of length two. It should be noted that a walk from v_i to v_ℓ has to occur immediately after that from v_j to v_i. Then, we define a time-aggregated preference matrix B^i by the (j,ℓ) element given by

$$B_{j\ell}^i = \sum_{t=2}^{t_{\max}} \frac{B_{j\ell}^i(t)}{\sum_{j',\ell'=1;j',\ell' \neq i}^{N} B_{j'\ell'}^i(t)}. \qquad (4.67)$$

A normalised betweenness preference matrix P^i is defined by the (j,ℓ) element given by

$$P_{j\ell}^i = \frac{B_{j\ell}^i}{\sum_{j',\ell'=1;j',\ell' \neq i}^{N} B_{j'\ell'}^i}, \qquad (4.68)$$

which is a probability of walks of length two from v_j to v_ℓ through v_i, for a given v_i.

The betweenness preference measure is defined using the mutual information by

$$\sum_{j,\ell} P_{j\ell}^i \log_2 \left[\frac{P_{j\ell}^i}{P^i(j)P^i(\ell)} \right], \qquad (4.69)$$

where $P^i(j) = \sum_{\ell'=1}^{N} P_{j\ell'}^i$ and $P^i(\ell) = \sum_{j'=1}^{N} P_{j'\ell}^i$. Empirical temporal networks tend to have statistically large betweenness preference values [Pfitzner *et al.* (2013)].

4.8   Null models and randomisation procedures

To know if a three-node subgraph is a significant network motif, we have to compare the abundance of the subgraph in a given network with that in a randomised network, or a null model (Section 3.9). Various random graph models are used as the null model. The situation is the same when we search for patterns in a given temporal network. For example, the temporal coherence is defined as the frequency of coherent triangles in a temporal network as compared to that when links are assumed to be activated independently (Section 4.4). The null model in this case is a temporal network in which links are present independently in time and of other links.

There are many null models of temporal networks, which preserve properties of the original temporal network to different extents. Many of them are based on randomisation of some elements in the original network. We list some examples below, assuming the event-based representation of temporal networks. For more null models, see [Holme and Saramäki (2012); Holme (2015)].

- Interval shuffling: For each link, randomly permute the inter-event times with the time of the first and last events fixed. The distribution of inter-event times is preserved. The structure of the aggregate network including the weight of each link is also preserved. Correlation and causality on each link and across different links are destroyed.
- Random times: For each link, redistribute the same number of events as that on the original link uniformly on $[0, t_{\max}]$. This procedure approximately corresponds to assigning an independent Poisson process to each link. As compared to the interval shuffling, the distribution of inter-event times is additionally destroyed. In various temporal network data, not all links are equally likely used in the entire observation period $[0, t_{\max}]$. Some links may be used earlier within $[0, t_{\max}]$, whereas others may be used later, reflecting an exit and entry of individual nodes. If we want to retain this property in the original data, we fix the first and last times of the events, denoted by t_0 and t_n, respectively, on each link and redistribute $n - 1$ events uniformly on $[t_0, t_n]$.

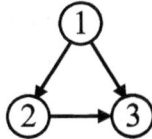

Fig. 4.13   A three-node subgraph in static directed networks.

- Link shuffling: Randomly shuffle the event sequence across links. The structure of the aggregate network as undirected network, distribution of link weight in the aggregate network and the set of event sequences are preserved. Correlation across links is destroyed.
- Random link shuffling: Replace a pair of randomly chosen links (v_i, v_j) and $(v_{i'}, v_{j'})$ by $(v_i, v_{j'})$ and $(v_j, v_{i'})$. The two event sequences on the original links are randomly given to the two new links. We repeat this procedure sufficiently many times. The distribution of inter-event times aggregated over links, distribution of the link weight and the degree of each node are preserved. The structure of the static network is destroyed.

4.9   Temporal motifs

Network motifs are overrepresented subgraphs in a network (Section 3.9). To avoid confusion, in this section we call them static network motifs and explain their extension to temporal networks, i.e., temporal network motifs (temporal motifs for short). To this end, we must be aware of the time and the order with which the links constituting a subgraph are used. Consider a three-node static network motif shown in Fig. 4.13, which is one of the 13 three-node subgraphs shown in Fig. 3.4. If this subgraph is significant in a static social network, it is likely that node v_1 sends information to nodes v_2 and v_3 then v_2 sends the information to v_3. Node v_3 may feel assured by confirming that v_2 says the same thing as what v_1 has said before. For this intuition to carry over to temporal networks, link (v_2, v_3) should be used after the other two links are used. In addition, if the three links are used in remote times, they are probably not related to each other in the temporal sense.

There are different definitions of temporal motifs that respect these requirements (see [Holme and Saramäki (2012); Kovanen *et al.* (2013a)] for surveys). Here we adhere to the definition by Kovanen and colleagues

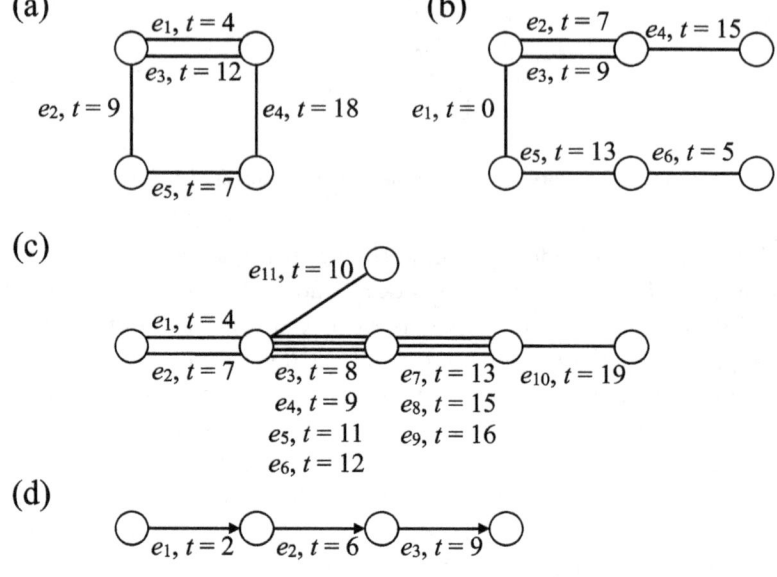

Fig. 4.14   Some temporal networks. A line represents an event.

[Kovanen *et al.* (2011, 2013a)]. For simplicity, we consider the event-based representation of temporal networks.

A temporal pattern, which is a candidate of temporal motif, should consist of a sequence of so-called Δt-adjacent events. Two events are defined to be Δt-adjacent if they occur within time difference Δt (but not simultaneously) and they share a node. In the undirected temporal network shown in Fig. 4.14(a), where a line represents an event, events e_1 and e_2 are Δt-adjacent when $\Delta t = 10$. Events e_1 and e_3, which occur on the same node pair, are also Δt-adjacent. Events e_1 and e_4 are not so because they are too far apart in time. e_1 and e_5 are not, either, because they do not share a node.

Two events are defined to be Δt-connected if there is a sequence of events connecting the two events such that each pair of events successive in time is Δt-adjacent. Consider the temporal network shown in Fig. 4.14(b). Events e_1 and e_4 are Δt-connected with $\Delta t = 10$ through e_2 or e_3. Events e_1 and e_6 are not Δt-connected although they occur within a time window of Δt. In fact, the concept of Δt-connectedness induces explosion in the counting of Δt-connected events. For example, e_1 and e_4 in Fig. 4.14(b) are Δt-connected according to either of the three sequences of events (e_1, e_2,

e_4), (e_1, e_3, e_4), or (e_1, e_2, e_3, e_4). An extreme case shown in Fig. 4.14(c) illustrates an explosion inherent in the counting; e_1 and e_{10} are Δt-connected in many different ways. The only condition required for the sequence of events connecting e_1 and e_{10} is that the sequence has to contain at least one of e_3, e_4, e_5 or e_6 and at least one of e_7, e_8 or e_9. In addition, a sequence of events Δ-connecting e_1 and e_{10} may or may not contain e_2 or e_{11}. Therefore, there are $2 \times (2^4 - 1) \times (2^3 - 1) \times 2 = 420$ sequences that Δt-connect e_1 and e_{10}.

Therefore, we focus on the valid temporal subgraph, which is defined as the sequence of Δt-adjacent events that do not skip any event whose start node is shared by an event that belongs to the sequence. For example, in Fig. 4.14(b), the valid temporal subgraph Δt-connecting e_1 and e_4 is given by (e_1, e_2, e_3, e_4). In Fig. 4.14(c), the valid temporal subgraph Δt-connecting e_1 and e_{10} is the one containing all the 11 events. It should be noted that e_{11}, which does not lie on the path from e_1 to e_{10} even in the aggregate network, is contained in the valid temporal subgraph. A subgraph of a valid temporal subgraph is qualified as valid temporal subgraph if the conditions are satisfied. For example, in the directed temporal network shown in Fig. 4.14(d), events e_1, e_2 and e_3 constitute the maximal valid temporal subgraph. Furthermore, the subgraph composed of just e_1 and e_2 are also a valid temporal subgraph. So are the subgraph composed of e_2 and e_3, and that composed of just e_1.

For algorithms to enumerate valid temporal subgraphs, we refer to [Kovanen *et al.* (2011, 2013a)]. A temporal motif is an equivalence class of valid temporal subgraphs. An equivalence class is defined by all valid temporal subgraphs with a given order of events, irrespectively of the precise event times. For example, temporal subgraphs $(v_1, v_2, t_1), (v_2, v_3, t_2)$ and $(v_4, v_5, t_3), (v_5, v_6, t_4)$ belong to the same equivalence class if $0 < t_2 - t_1, t_4 - t_3 < \Delta t$, even if $t_2 - t_1 \neq t_4 - t_3$. Similarly to the case of static network motifs, we can calculate the frequency of a valid temporal subgraph relative to that in a null model using the Z score (Eq. (3.72)). Some null models are explained in Section 4.8. More succinctly and also reliably, for the temporal subgraphs that differ only in the order of events, such as the two shown in Fig. 4.15, the simple comparison of the frequency counts tells which subgraph is more abundant than the other.

To compare temporal and static motifs, consider three-node directed static networks shown in Fig. 3.4. Corresponding to each of them, we can enumerate all possible temporal subgraphs. For example, subgraph 1 in Fig. 3.4 is associated with only one temporal subgraph (with multiple events

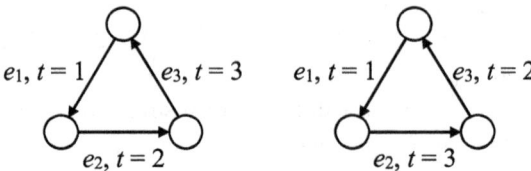

Fig. 4.15   Temporal subgraphs with three events that differ in the order of events.

Fig. 4.16   Two temporal subgraphs that are reduced to pattern 9 in Fig. 3.4 when the temporal information is discarded.

on a single link excluded for the discussion here). Subgraph 9 is associated with two temporal subgraphs shown in Fig. 4.16. However, not all temporal motifs have the corresponding static network motifs. Temporal subgraphs including repeated events do not have static counterparts because static network motifs usually (and implicitly) assume unweighted subgraphs. The undirected temporal subgraph shown in Fig. 4.15 is such an example.

The fact that a subgraph is a static motif in the aggregate network does not imply that an associated temporal subgraph is a significant temporal motif or vice versa. Obviously, in an extreme case in which all events occur far apart from each other in time, a static motif in the aggregate network never has a temporal counterpart. Conversely, because the temporal motif and static motif are defined based on different null models, a significant temporal motif does not induce a significant static network motif in the aggregate network.

Kovanen and colleagues measured temporal motifs in mobile phone records [Kovanen *et al.* (2011, 2013a,b)]. A phone call defines an event directed from one node to another. With $\Delta t = 10$ min, they found that some temporal subgraphs such as a repeated contact, returned contact, and some three-node patterns were significant temporal motifs. They mainly analysed two-node temporal motifs. In fact, in contrast to the case of static networks, there are a multitude of three-node temporal subgraphs because the same link can be repeatedly used in a temporal subgraph (see Fig. 4.15 for an example). They also carried out analysis by distinguishing node types, as classified by the gender, age and so on. In other words, isomorphic temporal subgraphs with different types of nodes are considered as different patterns. They found temporal homophily, that is, the same type of node tends to participate in an overrepresented temporal motif more often than they belong to a static network motif in the aggregated network. It should be noted that the distinction based on the node type can also be done for static network motifs.

4.10 Detection of change points and anomalies

It is often the case that a temporal network is non-stationary in the observation period. Many of the data analysis methods introduced so far implicitly assume that a given temporal network is stationary. Exceptions include detrended fluctuation analysis and time-dependent centrality measures. The generative models of temporal networks introduced in the next chapter also assume stationarity.

In fact, a temporal network may considerably change in time, often suddenly. Such changes are induced by faults in computer systems, abnormal physiological states in humans, frauds and spams, disease outbreaks, disturbances to an ecosystem and so forth. Methods to detect changes in general time series have been studied for much longer time than temporal networks in the fields of process control and statistics. These methods have been applied to temporal networks. In this section, we explain principles of these methods and basics of them, confining ourselves to statistical techniques. For more detailed exposure including various non-statistical methods, see [Chandola *et al.* (2009); Akoglu *et al.* (2015)].

Change-point and anomaly detection problems are usually formulated for the snapshot representation of temporal networks. In the change-point detection problem, we are interested in possible existence of a change in the nature of a given temporal network and its time t_c (Fig. 4.17(a)). By solving a change-point problem, we divide the evolution of a network into a succession of regimes in each of which the network structure is relatively stable. In a family of statistical methods of change-point detection, we resort to statistical hypothesis testing. The null hypothesis that no change point exists, denoted by H_0, is given by

$$H_0 : \Sigma(t) = \Sigma_0 \quad (1 \leq t \leq t_{\max}), \tag{4.70}$$

where $\Sigma(t)$ is a metric, or network footprint, which summarises snapshot

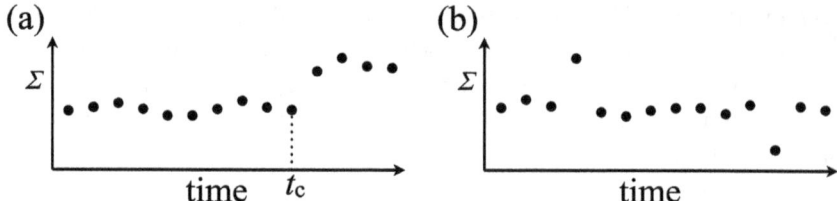

Fig. 4.17   Schematic of change-point and anomaly detection. In (a), metric Σ suddenly increases after $t = t_c$. In (b), there are apparently two anomalous snapshots.

$G(t)$. The change-point hypothesis, denoted by H_1, is given by

$$H_1 : \text{There exists } t_c \text{ such that } \Sigma(t) = \begin{cases} \Sigma_{\text{before}} & (1 \le t \le t_c), \\ \Sigma_{\text{after}} & (t_c + 1 \le t \le t_{\max}), \end{cases}$$

(4.71)

where $\Sigma_{\text{before}} \ne \Sigma_{\text{after}}$. The change point t_c can be determined by finding the value of t that maximises the discrepancy between the inferred Σ_{before} and Σ_{after}.

In the anomaly detection problem, we are interested in the time when the network behaves very differently from a typical time. Therefore, hypothesis H_1 should be altered to read that there exist some snapshots for which $\Sigma(t)$ is significantly different from the baseline value Σ_0. The snapshots with $\Sigma(t) \ne \Sigma_0$ are judged to be outliers. In practice, the normal behaviour itself can vary over time, which poses a challenge. Some methods incorporate trend estimation into the anomaly detection algorithm. Alternatively, we can regard that change-point detection itself aims at identifying significant shifts in the norm, or the trend.

In both change-point and anomaly detection, it is crucial to use a metric Σ that nicely summarises a snapshot. Any function of a network logically serves as a metric. Because different snapshots are usually not the same in the strictest sense, Σ should not overemphasise subtle differences between snapshots. At the same time, we do not want to miss "important" differences between snapshots.

The following list shows some simple and common choices for Σ [Pincombe (2005)]. Some of them are "delta" metrics, i.e., based on the difference between $G(t-1)$ and $G(t)$. When this type of Σ is used, H_0 is $\Sigma(t) = 0$ (i.e., no change), whereas H_1 is $\Sigma(t) \ne 0$ (i.e., change point or anomaly). It should be noted that relabelling of nodes and the corresponding relabelling of the links are allowed to make $G(t-1)$ and $G(t)$ the most similar in terms of Σ.

- Metrics for the static network that summarise an entire network. Examples include the average path length and clustering coefficient.
- Graph edit distance defined as the number of edits needed to transform one network to another. Denote the number of nodes and links in snapshot $G(t)$ by $N(G(t))$ and $M(G(t))$, respectively. The graph edit distance at time t is given by

$$N(G(t-1)) + N(G(t)) - 2N(G(t-1) \cap G(t))$$
$$+ M(G(t-1)) + M(G(t)) - 2M(G(t-1) \cap G(t)), \qquad (4.72)$$

where $G(t-1) \cap G(t)$ is the network composed of the nodes and links that commonly exist in $G(t-1)$ and $G(t)$.

- Metrics based on the maximum common subgraph. The maximum common subgraph of $G(t-1)$ and $G(t)$, denoted by $\mathrm{mcs}(G(t-1), G(t))$, is the induced subgraph of both $G(t-1)$ and $G(t)$ containing the largest number of nodes. By definition, any link (v_i, v_j) in $G(t-1)$ must be included in $\mathrm{mcs}(G(t-1), G(t))$ if both v_i and v_j are included in $\mathrm{mcs}(G(t-1), G(t))$. Such a link must also exist in $G(t)$. Several metrics have been developed on the basis of the maximum common subgraph. For example, the maximum common subgraph edge (i.e., link) distance is defined by

$$1 - \frac{M\left(\mathrm{mcs}(G(t-1), G(t))\right)}{\max\{M(G(t-1)), M(G(t))\}}. \qquad (4.73)$$

If we replace M (i.e., number of links) by N (i.e., number of nodes) in Eq. (4.73), we obtain the maximum common subgraph vertex (i.e., node) distance.

- Spectral distance, i.e.,

$$\sqrt{\frac{\sum_{i=1}^{n_{\mathrm{eig}}} \left[\lambda_{N+1-i}(G(t-1)) - \lambda_{N+1-i}(G(t))\right]^2}{\min\{\sum_{i=1}^{n_{\mathrm{eig}}} \left[\lambda_{N+1-i}(G(t-1))\right]^2, \sum_{i=1}^{n_{\mathrm{eig}}} \left[\lambda_{N+1-i}(G(t))\right]^2\}}}, \qquad (4.74)$$

where $\lambda_{N+1-i}(G(t))$ is the ith largest eigenvalue of the Laplacian of $G(t)$. n_{eig} is arbitrary, and there is no reason to exclude smallest Laplacian eigenvalues or the eigenvalues derived from other matrices such as the adjacency matrix.

- A parameter of a generative model inferred from $G(t)$. For pedagogical purposes, assume the Erdős-Rényi random graph for each snapshot and estimate the probability of a link, q. Then, the maximum likelihood estimator $\hat{q} = 2M(G(t))/\{N(G(t))[N(G(t))-1]\}$ is a qualified metric.

4.11   Link prediction

Predicting links that appear in the future is often beneficial. For example, if we know that a contact between a particular pair of individuals and at a particular time point will occur with a higher probability than otherwise, we may be able to intervene in the predicted pairs to mitigate the spread of infectious diseases or promote viral information spreading. In a collaboration network composed of academic authors, a successful link prediction implies who are likely to collaborate in a near future, given the pattern of collaborations in the past.

Link prediction is a question relevant to temporal networks in at least two ways. First, predicting the location of a future link implicitly assumes that networks vary over time. Growing network models (Section 3.8.3) and temporal network models introduced in the next chapter are also doing link prediction in a sense in particular when the models are informed by empirical data. Second, if we use the temporal information about the network, we may be able to predict links with a higher accuracy than when we regard the network to be static.

In this section, we explain the concept of link prediction and relatively simple methods, classified as similarity-based methods [Liben-Nowell and Kleinberg (2007); Hasan and Zaki (2011); Lü and Zhou (2011)]. For other methods based on maximum likelihood and machine learning, see other reviews [Hasan and Zaki (2011); Lü and Zhou (2011)].

Suppose a temporal network in either event-based or snapshot representation, observed in the time window $[0, t_{\max}]$. Given all time-stamped events or snapshots contained in time interval $[t_0, t_0']$, the objective is to exploit connectivity patterns observed in $[t_0, t_0']$ to predict links that will appear in time interval $[t_1, t_1']$, where $0 \leq t_0 < t_0' < t_1 < t_1'$. We refer to $[t_0, t_0']$ as the training interval and $[t_1, t_1']$ the test interval. To validate the accuracy of prediction, we must set $t_1' \leq t_{\max}$. If we do not do the validation, it suffices to respect $t_0' \leq t_{\max}$. We follow [Liben-Nowell and Kleinberg (2007)] to restrict the target of the prediction to the node pairs for which the link is not present in $[t_0, t_0']$. However, in temporal network applications, we are quite often interested in predicting a link in the test interval even if the link has appeared in the training interval.

A principled way to formulate the link prediction problem is to first assign a score to each node pair (v_i, v_j) based on the data in the training interval. The score determines the rank in terms of the likelihood with which the pair forms a link in the test interval.

Similarity-based methods determine the rank by assuming that similar nodes tend to get connected. Then, the problem of ranking node pairs is boiled down to designing of a similarity measure. Denote by $\mathcal{N}_i$ the set of neighbours of node v_i within the training period. Many similarity-based scores used in link prediction attach a high score to (v_i, v_j) when $\mathcal{N}_i$ and $\mathcal{N}_j$ have a large overlap. The common neighbours method defines the score by

$$|\mathcal{N}_i \cap \mathcal{N}_j|, \tag{4.75}$$

i.e., the number of common neighbours of v_i and v_j. The Jaccard index (Section 2.6) provides another scoring:

$$\frac{|\mathcal{N}_i \cap \mathcal{N}_j|}{|\mathcal{N}_i \cup \mathcal{N}_j|}. \tag{4.76}$$

The Adamic-Adar measure [Adamic and Adar (2003)] uses

$$\sum_{\ell \in \mathcal{N}_i \cap \mathcal{N}_j} \frac{1}{\log |\mathcal{N}_\ell|}. \tag{4.77}$$

These three similarity-based scores are based on the local neighbourhood of v_i and v_j. Another possibility is to apply the Katz centrality (Section 3.7.3), which considers all walks of different lengths between v_i and v_j and hence global structure of the network. The score based on the Katz centrality is defined by

$$\sum_{\ell=1} \alpha^\ell \times (\text{number of walks of length } \ell \text{ between } v_i \text{ and } v_j)$$
$$= [(I - \alpha A)^{-1} - I]_{ij}, \tag{4.78}$$

where α $(0 < \alpha < 1)$ is a discount factor for long walks.

For various data, the performance of these simple similarity-based scoring methods is comparable with that of more complicated ones [Liben-Nowell and Kleinberg (2007)]. Nevertheless, the accuracy of similarity-based and other algorithms is generally low. Link prediction is a difficult task.

The similarity-based algorithms introduced so far do not exploit the temporal structure of the data, as the network in the training interval is regarded as static. However, we can improve the accuracy by taking into account temporal information, for instance by giving more importance to recent links than old links [Lee *et al.* (2012)] or by enriching the similarity scores between nodes. In [Tabourier *et al.* (2016)], the authors focus on the problem of predicting links between neighbours of a focal node, i.e., ego, denoted by v. The only information at hand is the time series of activation of each of the k links incident to v. For static networks, a weighted and egocentric variant of the common neighbours similarity metric between the ith and jth neighbours of v is defined as

$$\frac{w_i w_j}{\sum_{\ell=1}^{k} w_\ell}, \tag{4.79}$$

where w_ℓ is the weight of the link connecting v and the ℓth neighbour of v. We expect the ith and the jth neighbours of v to interact if each of them often interacts with v. For temporal networks, the timing of an interaction provides additional information. For a temporal network in the snapshot representation, we define the similarity score as

$$\frac{w_i(t_1)w_j(t_1) + \cdots + w_i(t_{\max})w_j(t_{\max})}{\sum_{\ell=1}^{k} [w_\ell(t_1) + \cdots + w_\ell(t_{\max})]}, \tag{4.80}$$

which extends Eq. (4.79). Empirical validation on mobile phone data showed that Eq. (4.80) yielded a better prediction accuracy than Eq. (4.79) did.

4.12 Communities in temporal networks

As is the case of static networks, community detection is an expanding field of study for temporal networks. As compared to its static counterpart, the conceptualisation of communities for temporal networks needs extra consideration. First, a community in a temporal network may exist only in a limited range of time. It may represent a group of persons that have been active only around the time of a specific social event. Similarly, it may represent a group of neurons that function in synergy only when a subject carries out a certain task. Furthermore, a community in a temporal network can experience the following types of events [Palla *et al.* (2007); Greene *et al.* (2010)] (Fig. 4.18):

- Birth: A community may emerge at time t.
- Death: A community that has existed up to time $t-1$ may die at time t.
- Merging: Multiple communities may merge into one at time t.
- Splitting: A community may split into some communities at time t.
- Growth (expansion): A community may grow in its size (number of nodes) at time t.
- Contraction: A community may shrink in its size at time t.

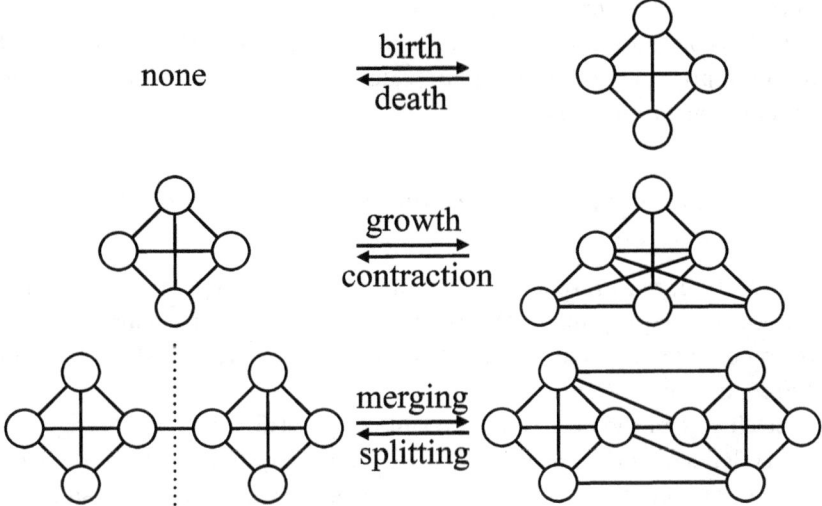

Fig. 4.18   Possible dynamic changes that a community in a temporal network experiences.

An algorithm for community detection in temporal networks should be able to tell when and how communities are involved in these events.

Identification of temporal communities is closely tied to change-point detection in temporal networks because a change in the organisation of communities at some point marks a change point. In addition, a minor change in the network structure may be better dismissed as being statistically non-significant, as is done in change-point/anomaly detection in general. In fact, some algorithms for community detection in temporal networks are founded on the framework of change-point detection.

Community detection in temporal networks is formulated in two different strands. A first type of algorithms aims at explicitly finding how communities evolve in time. Communities are typically detected in snapshots. The main problem is to determine, with a certain statistical accuracy, when and how a reorganisation of communities takes place. Some constraints on community structure in nearby snapshots may also be incorporated in order to ensure continuity of community structure in time. A second type of algorithms aims at finding one single community structure from the time series of networks regarding time t as a third dimension of the data, added to the two dimensions of the adjacency matrix. Again, communities are detected under some constraints such as continuity of a community in time.

In both cases, algorithms developed so far tend to operate on snapshot representations of temporal networks. Therefore, in this section we assume the snapshot representation and explain some algorithms of both types, some using modularity and others not.

4.12.1   *Modularity maximisation under estrangement constraint*

In this section, we present an evolving community type of algorithm. It maximises modularity under the explicit constraint that the partitioning results do not change too drastically between consecutive snapshots [Kawadia and Sreenivasan (2012)].

The change in the partitioning results is quantified by the estrangement, which is defined as the fraction of intra-community links at time $t - 1$ that become inter-community links at time t. The estrangement when snapshots are undirected networks are given by

$$E = \frac{1}{M(t)} \sum_{\substack{1 \leq i < j \leq N \text{ s.t.} \\ A_{ij}(t-1)=A_{ij}(t)=1}} \delta(g_i(t-1), g_j(t-1)) \left[1 - \delta(g_i(t), g_j(t))\right],$$

(4.81)

where $g_i(t)$ $(1 \leq i \leq N)$ is the label of the community that node v_i belongs to at time t, δ is Kronecker delta and $M(t)$ is the number of links in snapshot $G(t)$. According to Eq. (4.81), only links that are present at both times $t - 1$ and t are checked. For such a link (v_i, v_j), the contribution to the numerator is unity if and only if the link is an intra-community link at time $t-1$ and inter-community link at time t, i.e., estrangement. Otherwise, the contribution is equal to zero. See Fig. 4.19 for an example. Estrangement E ranges between zero and unity. If all the intra-community links at time $t - 1$ remain intra-community links or disappear at time t, $E = 0$. If all links are estranged, $E = 1$.

In each time step, we maximise modularity under the constraint that the estrangement is below a given threshold value. The threshold is a hyperparameter that we have to determine by other means. The problem is a standard optimisation problem under a constraint, which can be solved by various methods. Note that the method does not constrain on the links that have disappeared at t (which existed at time $t - 1$) or emerged at t.

Once the communities in $G(t)$ are determined, they are matched with the communities in $G(t - 1)$ via the maximisation of the Jaccard index (Eq. (2.91)). If community $\mathrm{CM}_c(t - 1)$ (i.e., cth community at time $t - 1$)

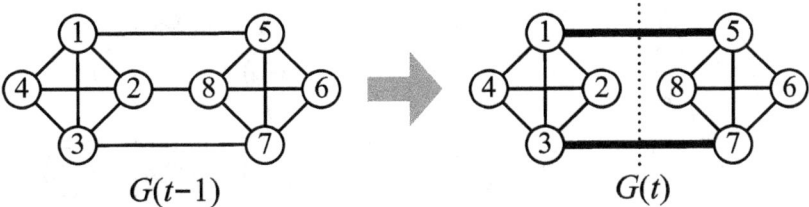

Fig. 4.19 Estrangement of links. If $G(t-1)$ has a single community and $G(t)$ has two communities, whose borders are shown by the dotted line, links (v_1, v_5) and (v_3, v_7), shown by the thick solid lines, are estranged links. Because there are 14 links in $G(t-1) \cap G(t)$, we obtain $E = 2/14$.

has the largest Jaccard index with $\text{CM}_{c'}(t)$ among all the communities at time t, and if $\text{CM}_{c'}(t)$ has the largest Jaccard index with $\text{CM}_c(t-1)$ among all the communities at time $t-1$, $\text{CM}_c(t-1)$ and $\text{CM}_{c'}(t)$ are matched as a single dynamic community. The communities at time t that do not find any matched partner at $t-1$ constitute new dynamic communities.

4.12.2 *Community matching approach*

Let us look at another evolving community type of algorithm based on community matching in different snapshots [Greene *et al.* (2010)].

Apply an arbitrary community detection algorithm for static networks to each snapshot. The method detects either disjoint or overlapping communities depending on the algorithm. The list of communities identified at time t, which we call step communities, is denoted by $\{C_{t,1}, \ldots, C_{t,N_{\text{CM}}(t)}\}$, where $N_{\text{CM}}(t)$ is the number of communities at time t. Then, the task is to match communities in different snapshots to regard that they belong to the same dynamic community denoted by C_c^{tp}.

In the example shown in Fig. 4.20, there are $t_{\max} = 3$ snapshots. There are five, seven and four step communities detected at $t = 1$, 2 and 3, respectively. By using the algorithm explained in the following, we conclude that there are seven dynamic communities in total in this example. Dynamic community C_1^{tp} is given by the sequence of step communities $\{C_{11}, C_{21}, C_{31}\}$. The three step communities C_{11}, C_{21}, and C_{31} are generally different from each other. However, they are required to be similar to comply with the thesis that a dynamic community should not change too drastically in a single time step. As another example, C_4^{tp} is formed by $\{C_{13}, C_{24}, C_{33}\}$.

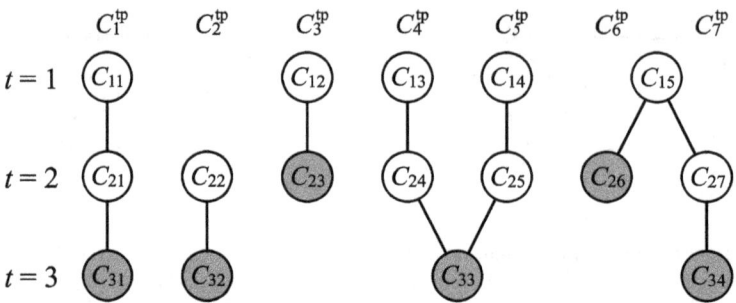

Fig. 4.20   Schematic of dynamic communities detected by the community matching algorithm [Greene *et al.* (2010)].

A step community may be shared by multiple dynamic communities (e.g., C_{15} and C_{33} in Fig. 4.20). This case corresponds to a merge or split. A dynamic community may not cover the entire observation period (e.g., C_2^{tp}, C_3^{tp} and C_6^{tp} in Fig. 4.20). Such a case corresponds to a birth or death. Expansion and contraction are identified by the difference between step communities belonging to the same dynamic community (e.g., difference between C_{11} and C_{21}). A node may belong to multiple dynamic communities even if the algorithm to detect step communities in a single snapshot forbids overlapping communities. For example, a node may belong to C_{11} and C_{33}. In this case, this node belongs to dynamic communities C_1^{tp}, C_4^{tp} and C_5^{tp}.

Step communities are matched as follows. We refer to the most recent observation for each dynamic community as the front of the dynamic community. The fronts of the seven dynamic communities at $t = 3$ are shown by the shaded circles in Fig. 4.20. Note that the fronts of C_3^{tp} and C_6^{tp} are step communities at $t = 2$ because they have at least temporarily dead at $t = 3$. However, it does not imply that the two dynamic communities ended their lives because they may revive in later time.

Then, we detect step communities at $t = 4$, i.e., $\{C_{41}, \ldots, C_{4N_{\mathrm{CM}}(4)}\}$. We attempt to match each of these communities with each front. To this end, we calculate the similarity between C_{4c} ($1 \leq c \leq N_{\mathrm{CM}}(4)$) and each front using the Jaccard index. For example, for C_{41}, we calculate $|C_{41} \cap C_{31}|/|C_{41} \cup C_{31}|$, $|C_{41} \cap C_{32}|/|C_{41} \cup C_{32}|$, $|C_{41} \cap C_{23}|/|C_{41} \cup C_{23}|$ and so on. If the Jaccard index exceeds a prescribed threshold θ ($0 \leq \theta \leq 1$), the pair is matched and C_{4c} is added to the dynamic community that the matched front belongs to. Then, we update the front of the corresponding dynamic community to C_{4c}. Step community C_{4c} may be matched to multiple fronts, representing a merger. In contrast, C_{4c} may not be matched to any of the existing dynamic community. In this case, we create a new dynamic community at $t = 4$ whose front is C_{4c}. We repeat this procedure by incrementing t.

This algorithm is of course not the only one for matching step communities. In an overlapping community detection method (Section 3.10.4) extended to the case of temporal networks [Palla *et al.* (2007)], matching between the communities at time t and those at $t + 1$ is done as follows.

First, we create the joint graph of $G(t)$ and $G(t + 1)$, denoted by $G(t) \cup G(t + 1)$. By definition, $G(t) \cup G(t + 1)$ contains all nodes and links that either $G(t)$ or $G(t + 1)$ possesses. Second, we apply the overlapping community detection algorithm [Palla *et al.* (2005)] to $G(t)$, $G(t + 1)$ and $G(t) \cup G(t + 1)$. By the nature of the method, any community in $G(t)$ is contained in exactly one community in $G(t) \cup G(t + 1)$. Likewise, any community in $G(t + 1)$ is contained in exactly one community in $G(t) \cup G(t + 1)$. Therefore, if a community in $G(t) \cup G(t + 1)$ contains just one community in $G(t)$ and just one in $G(t + 1)$, we match them. If a community in $G(t) \cup G(t + 1)$ contains multiple communities in either $G(t)$ or $G(t + 1)$, the communities are matched on the basis of the value of the relative overlap. The relative overlap is defined by the Jaccard index as $|C_c(t) \cap C_{c'}(t + 1)| \, / \, |C_c(t) \cup C_{c'}(t + 1)|$, where $C_c(t)$ is the cth community at time t that is contained in $G(t) \cup G(t + 1)$ and similar for $C_{c'}(t + 1)$. The cth community at time t and the c'th community at time $t + 1$ that maximise the relative overlap among all the possible c and c' pairs are matched. The other step communities at time t, corresponding to the different values of c, are interpreted to merge into this dynamic community. The other step communities at time $t + 1$, corresponding to the different values of c', are interpreted to split out of this dynamic community.

With this algorithm applied to empirical data, it was found that larger communities tended to have longer life time [Palla *et al.* (2007)].

4.12.3 *Mapping change*

In general, sensitivity to parameters and degeneracy are central problems when tracking evolving modularity. This problem is particularly relevant when we track temporal evolution of communities obtained by optimisation of a quality function whose landscape is rugged, as in the case of modularity or Infomap. In this case, a minor modification to the network may yield significantly different partitioning of the nodes. Even when the underlying network does not change, partitioning may substantially depend on trials if we use a non-deterministic community detection algorithm. An option for circumventing this problem and distinguishing meaningful trends from noise is to use bootstrapping techniques. In this section, we explain such a method for temporal networks, called the alluvial diagram [Rosvall and Bergstrom (2010)].

Each snapshot is assumed to be a weighted network. We proceed as follows:

(1) Partition each snapshot.
A method for detecting disjoint communities (i.e., to partition the set of nodes) in static networks is applied to each snapshot $G(t)$.

(2) Generate and partition an ensemble of bootstrap snapshot networks.
We resample a large number of bootstrap snapshot networks from each of the original snapshots. For each $G(t)$, we generate bootstrap replicates $G^*(t)$ by resampling the (integer-valued) weight of every link in $G(t)$ from a Poisson distribution whose mean is equal to the observed link weight. For unweighted networks, a different randomisation method should be used. We generate a large number of networks (i.e., $G^*(t)$) that fluctuate around $G(t)$. Then, we partition each $G^*(t)$ using the same method as in step (1). Because this step requires optimisation of the quality function many times (i.e., number of bootstrap networks $\times t_{\max}$), an efficient method such as the Louvain method is required.

(3) Determine the significance of the communities.
We find communities that are robust against fluctuations of the original network as follows. For each community in the original snapshot $G(t)$, we look for the largest subset of nodes that are also clustered together in at least 95% of the bootstrap partitions. Let us call this subset the core of the community. The size of a subset is measured in terms of either the number of nodes or the total flow of random walkers passing through the nodes, i.e., the sum of the PageRank over the nodes. The latter choice is coherent with flow-based community detection methods such as Infomap and Markov stability. To efficiently search the large space of subsets of nodes in each community and uncover the core, it is recommendable to use a standard heuristic such as simulated annealing. A community is considered to be significantly distinct from others if, in at least 95% of the bootstrap partitions, its core is not contained in a cluster that also contains another core. Two communities are mutually non-significant if their cores are clustered together in more than 5%

of the bootstrap partitions. Non-significant communities are grouped together as follows. First, the smallest non-significant community is merged with the community with which it is the most often clustered together in the bootstrap partitions. Then, the procedure is done for the next smallest community and so on.

(4) Generate an alluvial diagram to show changes between snapshots.

We have separately analysed different snapshots. To connect communities across time, we construct an alluvial diagram to highlight dynamic changes of communities. An example is shown in Fig. 4.21. Each significant community in $G(t)$ occupies a column in the diagram and is connected to the preceding and succeeding significant communities by stream fields. The size of a community is represented by the height in the diagram. The communities are ordered vertically by size, with mutually non-significant communities placed together and separated by a smaller spacing than a spacing between mutually significant communities. Darker colours represent nodes that are assigned to the core of a significant community, while lighter colours represent non-significant ones. Finally, stream fields connect different types of nodes in successive columns. By default, the thickness of the stream as well as that of blocks and communities is proportional to the total flow volume for the participating nodes. For clarity, we may hide the stream fields that are thinner than a threshold.

The resulting alluvial diagram visualises changes in the community structure in the form of flows linking the communities at different times. It also shows which events are statistically significant. In an example shown in Fig. 4.21, there are two major changes in the structure, but neither is significant. First, the split of the blue community leads to two significant communities, blue and purple, which are not mutually significant. Second, in the merger of the orange and red communities, the nodes in the core of the orange community are not transformed into the core of the red community.

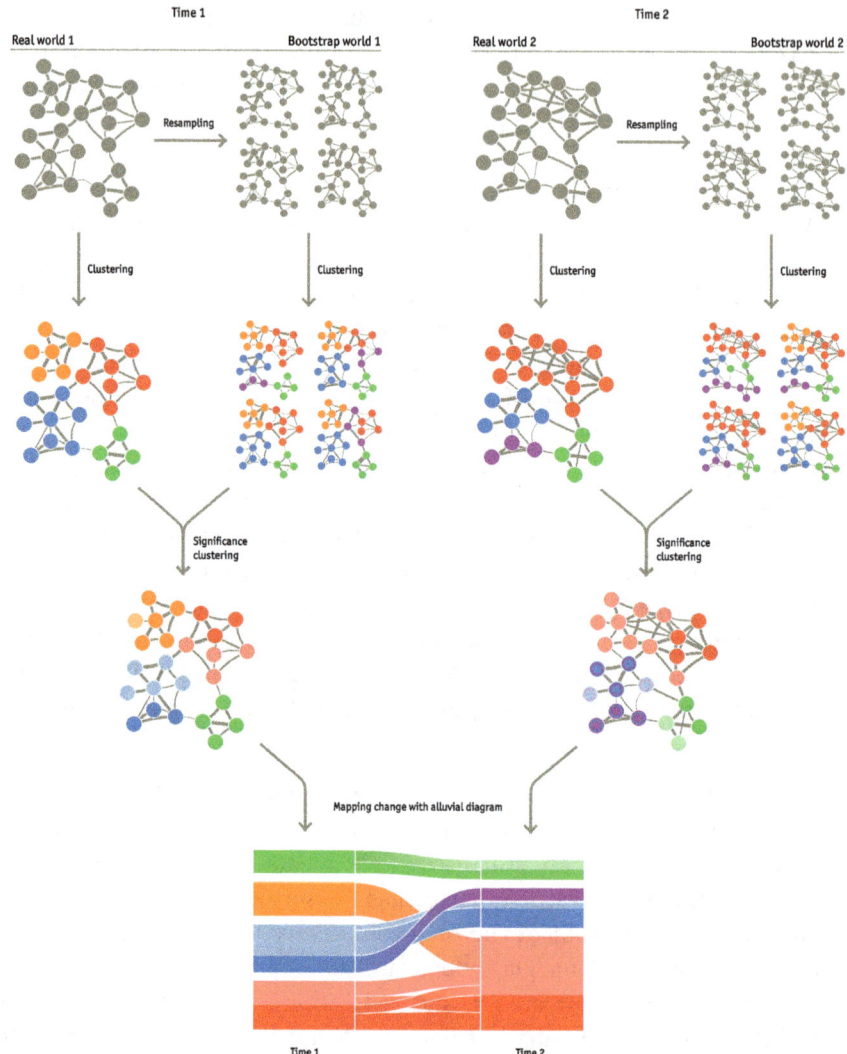

Fig. 4.21   Construction of an alluvial diagram. At each time, the original network and bootstrap networks are partitioned (clustered). The resulting communities are compared for identifying nodes belonging to the core of a community, as well as the significance of the communities. Darker colours indicate that a node belongs to the core of a significant community. For example, at $t = 1$, only 3 nodes belong to the core of the blue community. Two colours are used when two communities are not mutually significant. For instance, analysis shows that the blue and purple communities are not mutually significant at $t = 2$. Therefore, blue nodes are also identified as purple by being surrounded by purple and vice-versa. In the alluvial diagram, a smaller interval is used between the two communities. Stream fields connect the structure in successive times. Figure adapted from [Rosvall and Bergstrom (2010)] with permission. Also see the alluvial generator available at http://mapequation.org/apps/MapGenerator.html.

4.12.4 *Model-based approach*

All the methods explained so far are model-free methods. An alternative model-based approach to evolving community structure is to assume a generative stochastic model of temporal networks with dynamic community structure. Then, we infer the parameters of the model and interpret that the parameters that code communities in the inferred model represent dynamic community structure. Such attempts have been made with the temporal extensions of, for example, the stochastic block model [Yang *et al.* (2011); Xu and Hero III (2014); Peixoto and Rosvall (2015)] and a hierarchical random graph model [Peel and Clauset (2015)].

4.12.5 *Multilayer modularity*

A second type of approach to community detection aims at finding a decomposition of the network into communities in an extended network in which snapshots are concatenated. For example, we connect consecutive snapshots by auxiliary links to favour continuity of communities across time. Similarly to the first type of approach, an underlying assumption is that a community in a temporal network should be at least to some extent persistent over time and also may evolve. In the first type of approach, we found communities in each time window and compared their composition in successive steps. Instead, the idea here is to perform one single operation, such as modularity maximisation, on a larger, multilayer network equivalent to the original temporal network in the snapshot representation [Mucha *et al.* (2010); Bazzi *et al.* (2016)]. The multilayer framework has also been used for other applications in temporal networks, such as centrality [Taylor *et al.* (2015)].

Let us view a temporal network as a weighted network composed of $Nt_{\max}$ nodes by distinguishing each of the N nodes in different snapshots. In this extended network, a node is specified by a pair of indices (i, t), where $1 \leq i \leq N$ and $1 \leq t \leq t_{\max}$. The connectivity within each snapshot is given by $A(t)$. It is common to assume that inter-layer connections exist only between (i, t) and (j, t') for $1 \leq i, j \leq N$ symmetrically if and only if $i = j$ and $t' = t + 1$ ($1 \leq t \leq t_{\max} - 1$) with weight $\omega = C_{itt'} (= C_{it't}) (> 0)$. We set $C_{itt'} = 0$ if $t' \neq t \pm 1$. The multilayer network defined in this manner can be represented by either a three-way adjacency tensor or an extended adjacency matrix defined on $Nt_{\max}$ nodes.

Different versions of modularity can be defined on this tensor or matrix, depending on the interpretation given to inter-layer connections and on the null model. The simplest implementation is

$$\sum_{i',j'=1}^{Nt_{\max}} Q_{i'j'}^{\mathrm{mat}} \delta(g_{i'}, g_{j'}), \qquad (4.82)$$

where $g_{i'}$ is the community that the i'th node in the extended network with $Nt_{\max}$ nodes belongs to. The multilayer modularity matrix $Q^{\mathrm{mat}} = (Q_{i'j'}^{\mathrm{mat}})$ has the form

$$Q^{\mathrm{mat}} = \begin{bmatrix} Q^{\mathrm{mat}}(1) & \omega I & 0 & \cdots & & 0 \\ & \omega I & \ddots & \ddots & \ddots & \vdots \\ 0 & & \ddots & \ddots & \ddots & 0 \\ \vdots & & \ddots & \ddots & \ddots & \omega I \\ 0 & & \cdots & 0 & \omega I & Q^{\mathrm{mat}}(t_{\max}) \end{bmatrix}, \qquad (4.83)$$

where the $N \times N$ matrix $Q^{\mathrm{mat}}(t)$ is the modularity matrix (Eq. (3.73)) within layer t. The multilayer modularity given by Eq. (4.82) is rewritten as

$$\sum_{t=1}^{t_{\max}} \sum_{i,j=1}^{N} Q_{ijt}^{\mathrm{ten}} \delta(g_{i,t}, g_{j,t}) + 2\omega \sum_{t=1}^{t_{\max}-1} \sum_{i=1}^{N} \delta(g_{i,t}, g_{i,t+1}), \qquad (4.84)$$

where Q_{ijt}^{ten} denotes the (i,j)th entry of $Q^{\mathrm{mat}}(t)$. Equation (4.84) shows the different roles played by the two types of connections. Intra-layer connections determine the quality of clusters at a given time, and inter-layer connections favour continuity, or inertia, between layers. The parameter ω has an impact on temporal persistence of communities. In practice, the multilayer modularity maximisation can be heuristically solved with modified versions of standard modularity optimisation algorithms such as the Louvain algorithm.

To derive a different, popular expression of multilayer modularity for temporal networks, we extend the random-walk interpretation of modularity for static networks (Section 3.10.2) to the case of multilayer networks [Mucha *et al.* (2010)]. Now, random walkers are allowed to move from snapshot to snapshot as well as within each snapshot.

Consider the continuous-time random walk on the extended network with $Nt_{\max}$ nodes. Note that t, which represents the discrete time in the temporal network here, is not related to the continuous time for the random walk. To avoid confusions, we refer to the time for the random walk as $\tilde{t}$ in this section. The master equation for the random walk is given by

$$\frac{dp_{it}(\tilde{t})}{d\tilde{t}} = \sum_{t'=1}^{t_{\max}} \sum_{j=1}^{N} \frac{[A_{ij}(t)\delta_{tt'} + \delta_{ij}C_{jtt'}]\, p_{jt'}(\tilde{t})}{\kappa_{jt'}} - p_{it}(\tilde{t}), \tag{4.85}$$

where $\kappa_{jt'} = k_{jt'} + c_{jt'}$ is the weighted degree (i.e., strength) of the jth node in layer t', $k_{jt'} = \sum_{i=1}^{N} A_{ij}(t')$ is the intra-layer weighted degree of the same node and $c_{jt'} = \sum_{t''=1}^{t_{\max}} C_{jt't''}$ is the inter-layer one. The quantity in the summation on the right-hand side of Eq. (4.85) represents the rate at which the random walker moves from (j, t') to (i, t). Such a move to (i, t) is possible from the nodes in the same layer, (j, t), at rate $A_{ij}(t)/\kappa_{jt'}$ and from the ith node in the neighbouring layer, (i, t'), where $t' = t \pm 1$, at rate $C_{jtt'}/\kappa_{jt'}$. It should be noted that $C_{itt'} = 0$ if $t' \neq t \pm 1$. Under the assumption that each snapshot is an undirected network, the stationary state is given by

$$p_{it}^{*} = \frac{\kappa_{it}}{\sum_{t'=1}^{t_{\max}} \sum_{i'=1}^{N} \kappa_{i't'}} \equiv \frac{\kappa_{it}}{2\mu}. \tag{4.86}$$

By following the prescription for the static networks, we would like to calculate the probability that the random walker visits (j, t') at time $\tilde{t} = 0$ and (i, t) at time $\tilde{t} = \Delta\tilde{t}$. Within small time $\Delta\tilde{t}$, the walker initially at (j, t') may make a single jump to reach a node in the same layer, i.e., (i, t') $(1 \leq i \leq N)$ or a node in a neighbouring layer, i.e., (j, t) $(t = t' \pm 1)$. Under the independence assumption, which sets the null model, the walker moves on the configuration model network in the t'th layer or to (j, t) $(t = t' \pm 1)$ with weight $C_{jtt'}$ as in the original network. Then, the probability that walker visits (j, t') at time 0 and (i, t) at time $\Delta\tilde{t}$ under the independence assumption is given by

$$\left(\frac{k_{it}}{2M_t} \frac{k_{jt'}}{\kappa_{jt'}} \delta_{tt'} + \delta_{ij} \frac{C_{jtt'}}{c_{jt'}} \frac{c_{jt'}}{\kappa_{jt'}} \right) \frac{\kappa_{jt'}}{2\mu}, \tag{4.87}$$

where $2M_t = \sum_{j=1}^{N} k_{jt}$. In Eq. (4.87), $\kappa_{jt'}/2\mu$ is the probability that the random walker visits (j, t') at time 0 in stationarity. The quantity in parentheses represents the conditional probability that the walker visits (i, t) at time $\Delta\tilde{t}$ starting from (j, t') at time 0. A move occurs within the same layer t' with probability $k_{jt'}/\kappa_{jt'}$. If so, the walker moves to the ith node in the same layer with probability $k_{it'}/2M_{t'}$ according to the configuration model. This probability is represented by $(k_{it}/2M_t) \times \delta_{tt'}$. Alternatively, the walker moves to a neighbouring layer with probability $c_{jt'}/\kappa_{jt'}(= 1 - k_{jt'}/\kappa_{jt'})$. In this case, the destination is the jth node in layer $t = t' \pm 1$ with probability $C_{jtt'}/c_{jt'}$, which is accounted for by the second term in the parentheses.

Equation (4.85) indicates that the actual probability that the walker visits (j, t') at time 0 and (i, t) at small time $\Delta \tilde{t}$ is given by

$$\left[\delta_{ij}\delta_{tt'} + \Delta\tilde{t} \left(\frac{A_{ij}(t)\delta_{tt'} + \delta_{ij}C_{jtt'}}{\kappa_{jt'}} - \delta_{ij}\delta_{tt'} \right) \right] \frac{\kappa_{jt'}}{2\mu}. \qquad (4.88)$$

By subtracting Eq. (4.87) from Eq. (4.88) and taking the summation over the (i, t) and (j, t') nodes that belong to the same community in the extended network, we obtain

$$Q = \frac{1}{2\mu} \sum_{i,j,t,t'} \left[(1 - \Delta\tilde{t})\delta_{ij}\delta_{tt'} + \Delta\tilde{t}A_{ij}(t)\delta_{tt'} - \frac{k_{it}k_{jt'}}{2M_t}\delta_{tt'} + (\Delta\tilde{t} - 1)\delta_{ij}C_{jtt'} \right]$$

$$\times \delta(g_{it}, g_{jt'}). \qquad (4.89)$$

Because $\sum_{i,j,t,t'} \delta_{ij}\delta_{tt'}\delta(g_{it}, g_{jt'}) = Nt_{\max}$ is independent of the partitioning of the network, we ignore the first term on the right-hand side of Eq. (4.89). By rescaling $C_{jtt'}$ by a multiplicative factor of $(\Delta\tilde{t} - 1)/\Delta\tilde{t}$, we also ignore $(\Delta\tilde{t} - 1)$ in the fourth term. By allowing $\gamma \equiv 1/\Delta\tilde{t}$ to depend on the layer, corresponding to different diffusion time scales introduced to different layers, we write the final form of the so-called multilayer modularity as

$$Q = \frac{1}{2\mu} \sum_{i,j,t,t'} \left[A_{ij}(t) - \gamma(t)\frac{k_{it}k_{jt'}}{2M_t}\delta_{tt'} + \delta_{ij}C_{jtt'} \right] \delta(g_{it}, g_{jt'}). \qquad (4.90)$$

In fact, the summation over t' is restricted to $t' = t \pm 1$. Equation (4.90) is a special case of Eq. (4.84). In practice, when the jth node is absent in the tth snapshot, we set $C_{j,t,t-1} = C_{j,t,t+1} = 0$.

We have to choose a temporal resolution parameter, $\omega = C_{i,t,t+1}$ ($1 \leq i \leq N$, $1 \leq t \leq t_{\max} - 1$). If $\omega = 0$, different snapshots are regarded as independent static networks. If ω is very large, each node is assigned to the same community across all snapshots because the third term on the right-hand side of Eq. (4.90) is dominant. In the limit $\omega \to \infty$, however, Q is not the same as that for the aggregate network. This is because the second term on the right-hand side of Eq. (4.90) indicates that the configuration model for each layer, not for the aggregate network, is used as a part of the null model. In particular, for large ω, the optimal community assignment for each layer results from the single-layer maximisation problem on matrix $\sum_{t=1}^{t_{\max}} Q^{\mathrm{mat}}(t)$, i.e., the mean modularity matrix [Bazzi *et al.* (2016)]. In other words, $\sum_{i,j=1}^{N} \left(\sum_{t=1}^{t_{\max}} Q^{\mathrm{mat}}(t) \right)_{ij} \delta(g_i, g_j)$ is maximised. The third term on the right-hand side of Eq. (4.90) increases as ω increases more when many $\delta(g_{it}, g_{jt'})$ values are unity than when few of them are unity. Therefore, a larger ω value tends to yield fewer communities. In contrast, a larger $\gamma(t)$ value generally yields a larger number of communities.

Although we have focused on the null model in which each snapshot is a configuration model, there are many other possibilities for null models in the case of temporal networks [Bassett *et al.* (2013)]. For example, a temporal null model consists in randomly permuting the order of snapshots. This null model has been used for assessing the significance of temporal evolution of community structure [Bassett *et al.* (2013)].

4.12.6 *Tensor factorisation approach*

In this section, we explain a method using tensor factorisation, which is quite distinct from the community detection methods for temporal networks presented in the previous sections [Gauvin *et al.* (2014)]. The vector and matrix have one and two running indices, respectively, and are amenable to linear algebra tools. Tensor extends these concepts to the case of a larger number of indices. In particular, three-way tensors, which have also appeared in the previous section, are a natural representation of temporal networks. Let $A^{\text{ten}} = (A^{\text{ten}}_{ijt}) \in \mathbb{R}^{N \times N \times t_{\max}}$ be a three-way tensor whose first two indices range from 1 to N and the third index ranges from 1 to $t_{\max}$. Tensor A^{ten} is equivalent to the snapshot representation with the relationship $A^{\text{ten}}_{ijt} = A_{ij}(t)$.

Linear algebra tells us that a matrix, corresponding to a static network, allows various types of factorisation. In particular, it allows the singular value decomposition. If we only retain the most dominant modes in the singular value decomposition of a network, we obtain a low-rank approximation to the network, which represents a scaffold of the given network. A tensor allows similar decompositions and low-rank approximations. The idea is to use them and interpret the results as partitioning of a temporal network.

Consider the following canonical decomposition (also called by various other names [Kolda and Bader (2009)]) of the three-way tensor $A^{\text{ten}} \in \mathbb{R}^{N \times N \times t_{\max}}$. It represents A^{ten} as a sum of rank-one tensors as follows:

$$A^{\text{ten}}_{ijt} = \sum_{r=1}^{R_S} B_{ir} B'_{jr} B''_{tr}. \qquad (4.91)$$

The smallest value of R_S for which Eq. (4.91) holds true for $1 \le i, j \le N$

and $1 \leq t \leq t_{\max}$ defines the rank of tensor A^{ten}. For undirected temporal networks, $A^{\text{ten}}_{ijt} = A^{\text{ten}}_{jit}$, and hence we can enforce $B_{ir} = B'_{ir}$ ($1 \leq i \leq N$, $1 \leq r \leq R_S$). It should be noted that calculation of the rank is not a straightforward exercise for tensors in contrast to the case of matrices; it is an NP-hard problem [Kolda and Bader (2009)].

We can transform a three-way tensor to a matrix (i.e., two-way tensor) without losing information via an operation called the matricisation. To matricise A^{ten}, we use one of the three indices as the row index of the resulting matrix. The column index of the matrix is defined by unfolding the other two indices in the lexicographical order. Consider an example with $N = 2$ nodes and $t_{\max} = 3$ snapshots. When we keep the first index, the unfolding as a result of matricisation is given by

$$\mathbf{S}_{(1)} = \begin{pmatrix} A^{\text{ten}}_{111} & A^{\text{ten}}_{112} & A^{\text{ten}}_{113} & A^{\text{ten}}_{121} & A^{\text{ten}}_{122} & A^{\text{ten}}_{123} \\ A^{\text{ten}}_{211} & A^{\text{ten}}_{212} & A^{\text{ten}}_{213} & A^{\text{ten}}_{221} & A^{\text{ten}}_{222} & A^{\text{ten}}_{223} \end{pmatrix}. \tag{4.92}$$

If we keep the second or third index, we respectively obtain different matricisation results as follows:

$$\mathbf{S}_{(2)} = \begin{pmatrix} A^{\text{ten}}_{111} & A^{\text{ten}}_{112} & A^{\text{ten}}_{113} & A^{\text{ten}}_{211} & A^{\text{ten}}_{212} & A^{\text{ten}}_{213} \\ A^{\text{ten}}_{121} & A^{\text{ten}}_{122} & A^{\text{ten}}_{123} & A^{\text{ten}}_{221} & A^{\text{ten}}_{222} & A^{\text{ten}}_{223} \end{pmatrix}, \tag{4.93}$$

$$\mathbf{S}_{(3)} = \begin{pmatrix} A^{\text{ten}}_{111} & A^{\text{ten}}_{121} & A^{\text{ten}}_{211} & A^{\text{ten}}_{221} \\ A^{\text{ten}}_{112} & A^{\text{ten}}_{122} & A^{\text{ten}}_{212} & A^{\text{ten}}_{222} \\ A^{\text{ten}}_{113} & A^{\text{ten}}_{123} & A^{\text{ten}}_{213} & A^{\text{ten}}_{223} \end{pmatrix}. \tag{4.94}$$

In terms of the factor matrices $B \equiv (B_{ir}) \in \mathbb{R}^{N \times R_S}$, $B' \equiv (B'_{jr}) \in \mathbb{R}^{N \times R_S}$, and $B'' \equiv (B''_{tr}) \in \mathbb{R}^{t_{\max} \times R_S}$, the matricisation can be rewritten as

$$\mathbf{S}_{(1)} = B(B'' \odot B')^{\top}, \tag{4.95}$$

$$\mathbf{S}_{(2)} = B'(B'' \odot B)^{\top}, \tag{4.96}$$

$$\mathbf{S}_{(3)} = B''(B' \odot B)^{\top}, \tag{4.97}$$

where $\odot$ denotes the Khatri-Rao product, which is a column-wise Kronecker product of matrices. For example,

$$
B'' \odot B' = \left[\begin{pmatrix} B''_{11} \\ \vdots \\ B''_{t_{\max}1} \end{pmatrix} \otimes \begin{pmatrix} B'_{11} \\ \vdots \\ B'_{N1} \end{pmatrix} \begin{pmatrix} B''_{12} \\ \vdots \\ B''_{t_{\max}2} \end{pmatrix} \otimes \begin{pmatrix} B'_{12} \\ \vdots \\ B'_{N2} \end{pmatrix} \right.
$$

$$
\left. \cdots \begin{pmatrix} B''_{1R_S} \\ \vdots \\ B''_{t_{\max}R_S} \end{pmatrix} \otimes \begin{pmatrix} B'_{1R_S} \\ \vdots \\ B'_{NR_S} \end{pmatrix} \right]
$$

$$
= \begin{pmatrix} B''_{11}B'_{11} & B''_{12}B'_{12} & \cdots & B''_{1R_S}B'_{1R_S} \\ B''_{11}B'_{21} & B''_{12}B'_{22} & \cdots & B''_{1R_S}B'_{2R_S} \\ \vdots & \vdots & & \vdots \\ B''_{21}B'_{11} & B''_{22}B'_{12} & \cdots & B''_{2R_S}B'_{1R_S} \\ \\ \vdots & \vdots & & \vdots \\ \\ B''_{t_{\max}1}B'_{N1} & B''_{t_{\max}2}B'_{N2} & \cdots & B''_{t_{\max}R_S}B'_{NR_S} \end{pmatrix}, \tag{4.98}
$$

where $\otimes$ denotes the Kronecker product.

To obtain a low-rank approximation to the temporal network, A^{ten}, we look for low-rank matrices $\tilde{B}$, $\tilde{B}'$, $\tilde{B}''$ such that $\tilde{B}(\tilde{B}'' \odot \tilde{B}')^\top$, $\tilde{B}'(\tilde{B}'' \odot \tilde{B})^\top$ and $\tilde{B}''(\tilde{B}' \odot \tilde{B})^\top$ accurately approximate $\mathbf{S}_{(1)}$, $\mathbf{S}_{(2)}$ and $\mathbf{S}_{(3)}$, respectively. In particular, by imposing $\tilde{B}, \tilde{B}' \in \mathbb{R}^{N \times R}$, $\tilde{B}'' \in \mathbb{R}^{t_{\max} \times R}$, where $R < R_S$, we pursue a rank-R approximation to A^{ten}. Usually, we set $R \ll R_S$.

Let us now consider a variant of this decomposition by constraining the entries of B_{ir}, B'_{jr} and B''_{tr} in Eq. (4.91) to be non-negative. Then, the value of B_{ir} is interpreted as an extent to which the ith node participates in the rth mode. Computationally, the low-rank approximation is formulated as an optimisation problem, which can be solved by an alternating least squares algorithm [Gauvin *et al.* (2014)].

Our interest is community detection, not low-rank approximation to networks. How shall we interpret the results? The optimisation yields approximate factor matrices $\tilde{B} = (\tilde{B}_{ir})$, $\tilde{B}' = (\tilde{B}'_{jr})$ and $\tilde{B}'' = (\tilde{B}''_{tr})$. We regard that each rank-one tensor $\tilde{B}_{ir}\tilde{B}'_{jr}\tilde{B}''_{tr}$ $(1 \le r \le R)$ represents the rth community. Matrices $\tilde{B}$ and $\tilde{B}'$ $(\tilde{B} = \tilde{B}'$ if the network is undirected) encode affiliation of nodes to different communities, with $\tilde{B}_{ir}$ and $\tilde{B}'_{ir}$ quantifying how intensively the ith node belongs to the rth community. Matrix $\tilde{B}''$ encodes the temporal activity of the communities, with $\tilde{B}''_{tr}$ quantifying how intensively the rth community is active at time t. The strength of the

rth community at time t is defined by $B''_{tr} \sum_{i=1}^{N} B_{ir}$ for undirected temporal networks.

Because factor matrices $\tilde{B}$ and $\tilde{B}'$ are not binary, a node may participate in multiple communities to different extents. By exploiting this feature, the tensor factorisation approach allows the detection of overlapping communities. It should also be noted that, unlike the methods explained in the previous sections, the present method does not impose continuity of the results with respect to time. Therefore, a single community may be strong in early times, then disappear (as indicated by a small value of the strength of the rth community defined above), and reappear in later time. This property may not be desirable. Extending the method to account for temporal contiguity of communities seems to be an interesting question.

Chapter 5

Models of temporal networks

In this chapter, we review models for generating temporal networks.

5.1   Models of non-Markovianity

When modelling evolution of temporal networks or dynamical processes on networks, one is confronted with different types of non-Markovianity, or memory, as we will see in several occasions in this and next chapters. Here we clearly distinguish the following types of mechanisms leading to non-Markovianity.

(1) *Temporal non-Markov*: It is the norm rather than the exception that inter-event times obey long-tailed distributions, producing bursty behaviour (Section 4.6). When a generative model of or a dynamical process on temporal networks is modelled by a non-Poissonian renewal process, it is not Markovian. For example, in continuous-time random walks in which jumps are triggered by a non-Poissonian renewal process, trajectories are Markovian and can be reproduced by a classical Markov chain. However, predicting the location of a walker at a given time is a non-Markovian enquiry. We encounter this type of non-Markovianity with generative models of temporal networks (Sections 5.2 and 5.4), analysis tools for dynamics on temporal networks (Sections 6.1 and 6.2), random walks (Section 6.3) and epidemic processes (Section 6.4).

(2) *Pathway non-Markov*: A second type of non-Markovianity emerges when trajectories of a diffusive entity are not reproduced by a Markov chain. When you have received information from your acquaintance, where will it go to next? If modelled by a Markov process, any of your friends is equally likely to receive the information if the network

is unweighted. The probability would be proportional to the weight of the link if the network is weighted. In many empirical data, however, the next destination of the information often depends on where the information has come from. If you receive a news from your mother, it will be more likely to be communicated to another family member than to one of your professional colleagues. This type of memory can be modelled by higher-order Markov processes, where a certain number of steps in the past is used to determine the next step (Section 5.7). Non-Markovian pathways may emerge due to correlations between successive activations of links (Section 4.7).

(3) *History-dependent infection*: In standard models of epidemic spreading, a susceptible node gets infected at a constant rate, regardless of the amount of exposure to infected neighbours in the past. This modelling framework is appropriate for a majority of infectious diseases. However, in the case of information propagation, the exposure to information may increase an interest and trust in a topic, and the probability of being infected in the future may increase upon each attempt [Centola and Macy (2007)]. Such a process is sometimes called complex diffusion and is reproduced by, for example, threshold models [Granovetter (1978); Dodds and Watts (2004)] and general cascade models [Kleinberg (2007)]. Complex diffusion depends on its history. Because this mechanism is not specific to temporal networks, its coverage is limited in this book; we explain self-exciting processes, mostly as a single node property, in Section 5.5. In fact, models of complex diffusion may exhibit non-trivial properties when taking place on temporal networks, because bursts of activity may affect the probability for a node to reach a certain threshold in a short time window [Karimi and Holme (2013); Takaguchi *et al.* (2013); Backlund *et al.* (2014)].

5.2   Stochastic temporal networks

As a primer to various models for temporal networks, in this section, we consider a simple model. Consider an arbitrary static network on which a temporal network in the event-based representation is built. On each link, we randomly assign an event sequence using a renewal process with the distribution of inter-event times, $\psi(\tau)$. We call this model the stochastic temporal network model.

If $\psi(\tau)$ is the exponential distribution, the model is not really temporal for the following reason. Usually, events of stochastic processes, such as movements of random walkers and infection and recovery events in epidemic dynamics, occur as a Poisson process with an appropriate rate. Therefore, when we simulate stochastic dynamics such as the random walk and epidemic processes, we implicitly run a Poisson process with an appropriate rate on each link. Therefore, we do not regard that a generated network is temporal when $\psi(\tau)$ is the exponential distribution.

Because the inter-event time typically obeys a long-tailed distribution (Section 4.6.1), let us assume that $\psi(\tau)$ attached to each link is such a distribution. If we run a dynamical process on this temporal network, we can inspect the effect of long-tailed distribution of inter-event times on the dynamics, for example, whether random walkers tend to travel faster or slower, or whether the epidemic spreading is likely to occur on a large scale or not, as compared to the static-network case (i.e., with the exponential $\psi(\tau)$).

Extensions of the stochastic temporal network model include assigning different distributions of inter-event times to different links. We may also introduce correlation in τ by assuming a conditional distribution of the inter-event times given the previous inter-event time τ_{prev}, i.e., $\psi(\tau|\tau_{\text{prev}})$. These techniques are also useful in creating null models of temporal networks (Section 4.8).

5.3 Activity driven model

The activity driven model is a generative model of temporal networks [Perra *et al.* (2012b)]. This model generates temporal networks in the snapshot representation as follows.

Each node v_i ($1 \le i \le N$) is assigned with a random variable a_i, called activity potential, drawn from a given distribution $F(a)(a \ge 0)$. Activity potential a_i represents the tendency for v_i to form links in each snapshot and is fixed in time once it has been drawn. For static networks, assigning a node weight to each node and generating links as a function of the node weight is a common approach (e.g., [Goh *et al.* (2001); Caldarelli *et al.* (2002)]), akin to the configuration model.

In each snapshot, v_i ($1 \le i \le N$) is active with probability $a_i \Delta t$ and inactive with probability $1 - a_i \Delta t$, where Δt is the time difference between a consecutive pair of snapshots. We select $F(a)$ such that $a_i \Delta t$ never exceeds

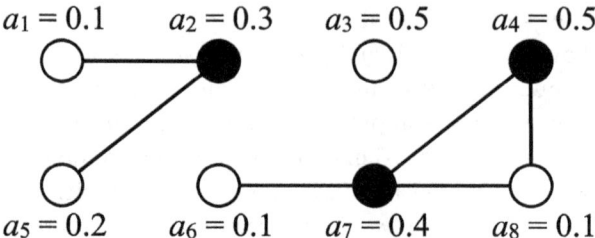

Fig. 5.1   A snapshot in the activity driven model. We set $N = 8$, $m = 2$ and $\Delta t = 1$. Filled circles represent active nodes. Open circles represent inactive nodes.

unity. Each activated node generates m undirected links that connect to randomly selected m other nodes (Fig. 5.1). Two nodes v_i and v_j can be simultaneously active and send links to each other. The treatment of this case is often implicit, but we can assign just one link between v_i and v_j or redirect one of the two links to a different destination node that is randomly drawn. Inactive nodes can also have links because they may receive links from active nodes. In this way, a snapshot, which is an undirected and unweighted network, is generated. To generate the next snapshot, we discard all the information about the previous snapshot except that we continue to use the same a_i $(1 \leq i \leq N)$. We repeat the procedure to generate a sequence of snapshots. The time to the next activation of v_i obeys the geometric distribution with success probability $a_i \Delta t$. In the limit $\Delta t \to 0$, the node activation occurs according to a Poisson process.

In a different variant of the model published earlier, activated nodes in each snapshot are randomly paired to form an event [Rocha *et al.* (2011)]. If the number of activated nodes is odd in a snapshot, a single remaining node stays isolated in this snapshot and will connect to an activated node available at the earliest time afterwards. In this model, inactive nodes do not receive links, and a snapshot is a collection of disjoint links.

The activity driven model is a Markovian model, in which different snapshots are independent of each other. Therefore, it does not explain temporal correlation inherent in various temporal network data. The model does not produce a long-tailed distribution of inter-event times, either. Because of the Markovian property, the inter-event time on each link, measured as the time between two consecutive appearances of the link, obeys a geometric distribution (i.e., no long tail).

The activity driven model can produce temporal networks in the event-based representation in the limit $\Delta t \to 0$. However, here we set $\Delta t = 1$, and therefore impose $0 \leq a_i \leq 1$ $(1 \leq i \leq N)$ [Starnini and Pastor-Satorras (2013)] to make the model fully compatible with the snapshot representation.

Despite the Markovian property, the activity driven model has been a useful tool since its inception. This is because the model is simple enough to facilitate analytical investigations of various structural and dynamical properties, while it produces a certain amount of temporality and heterogeneity of networks. Because we currently do not have standard models of temporal networks, like the configuration model (Section 3.8.2) and the Barabási-Albert model (Section 3.8.3) for static networks, such an analytically tractable model is an important asset. Structural properties [Perra *et al.* (2012b); Starnini and Pastor-Satorras (2013)], equilibrium properties of random walks [Perra *et al.* (2012a); Ribeiro *et al.* (2013)] and models of epidemic dynamics [Perra *et al.* (2012b); Liu *et al.* (2013, 2014); Rizzo *et al.* (2014); Starnini and Pastor-Satorras (2014)] have been analytically understood for the activity driven model. The simplicity of the model would also aid numerical simulations and interpretation of the results.

In each snapshot, the mean degree is equal to

$$\langle k \rangle \approx 2m\langle a \rangle, \tag{5.1}$$

where $\langle a \rangle = \int aF(a)\mathrm{d}a$ is the mean activity. We obtain Eq. (5.1) because there are $N\langle a \rangle$ activated nodes on an average, one activated node contributes m links, each of which contributes to the degree sum by 2, and hence to the mean degree by $2/N$.

Other structural properties of the activity driven model have been derived mostly for aggregate networks [Perra *et al.* (2012b); Starnini and Pastor-Satorras (2013)]. If we take sufficiently many snapshots for aggregation, the complete graph is the trivial outcome. Therefore, in the following let us focus on the aggregate network normalised by the number of snapshots $t_{\max}$.

Given a_i $(1 \leq i \leq N)$, the aggregate network normalised by the unit time is given by

$$A_{ij}^* = \frac{m\,(a_i + a_j)}{N} \tag{5.2}$$

up to the order of $1/N$, i.e., if we neglect the probability that an active node sends a link to another active node. Equation (5.2) implies that the mean degree of v_i in the aggregate network per unit time, denoted by k_i^*, is given by

$$k_i^* = \sum_{j=1}^{N} A_{ij}^* = m\,(a_i + \langle a \rangle). \tag{5.3}$$

Equation (5.3) indicates a monotonic relationship between k_i^* and a_i, with $dk_i^*/da_i = m$. Substitution of this relationship and Eq. (5.3) in $p(k^*)dk^* = F(a)da$ yields [Perra *et al.* (2012b); Starnini and Pastor-Satorras (2013)]

$$p(k^*) \approx \frac{1}{m} F\left(\frac{k}{m} - \langle a \rangle\right). \tag{5.4}$$

Therefore, the degree distribution of the aggregate network is essentially the same as the distribution of the activity, $F(a)$, up to a rescaling factor m. This result is intuitive because m is the number of links that an active node generates. Essentially the same relationship holds true for the variant of the model mentioned before [Rocha *et al.* (2011)]. A power-law degree distribution for the aggregate network is produced by a power-law $F(a)$. For this reason, a common choice is $F(a) = (\gamma - 1)\epsilon^{\gamma-1}a^{-\gamma}$, where $\epsilon \leq a \leq 1$, $\gamma > 2$ and $0 < \epsilon \ll 1$ to avoid divergence of the density of a around zero. Then, $p(k^*)$ is a power-law distribution with power-law exponent γ. Other properties include the following [Starnini and Pastor-Satorras (2013)]. First, the aggregate network has mildly negative degree correlation unless $F(a)$ is the delta function. Second, the clustering coefficient for the aggregate network is small.

By combining Eqs. (5.2) and (5.3), we obtain

$$A_{ij}^* = \frac{k_i^* + k_j^* - 2m\langle a \rangle}{N}. \tag{5.5}$$

Therefore, the expected degree affects the connectivity additively in the activity driven model. For the configuration model, we obtain $A_{ij}^* = k_i k_j / 2M$ (Eq. (3.64)), where we remind that M is the number of links in the network. Therefore, the expected degree affects A_{ij} multiplicatively. As aggregate networks, two random network models are different, although both models can generate an (almost) arbitrary degree distribution. It should be noted that the configuration model has no degree correlation, whereas the activity driven model does.

The eigenspectrum of the adjacency matrix of the aggregated network of the activity driven model, A^*, is also known [Speidel *et al.* (2016)]. The eigenequation is given by

$$A^* \boldsymbol{u} = \lambda \boldsymbol{u}, \tag{5.6}$$

where λ is an eigenvalue of A^* and $\boldsymbol{u} = (u_1, \ldots, u_N)^\top$ is the corresponding eigenvector. Assume that $\sum_{i=1}^N u_i = 1$, which is always possible by normalisation unless $\sum_{i=1}^{N} u_i = 0$. By summing

$$\sum_{j=1}^{N} A_{ij}^* u_j = \lambda u_i \tag{5.7}$$

over i, where A_{ij}^* is given by Eq. (5.2), we obtain

$$m \left(\frac{\sum_{i=1}^{N} a_i}{N} + \sum_{i=1}^{N} a_i u_i \right) = \lambda. \tag{5.8}$$

By multiplying a_i to both sides of Eq. (5.7) and summing over i, we obtain

$$m \left(\frac{\sum_{i=1}^{N} u_i^2}{N} + \frac{\sum_{i=1}^{N} a_i}{N} \sum_{i=1}^{N} a_i u_i \right) = \lambda \sum_{i=1}^{N} a_i u_i. \tag{5.9}$$

By combining Eqs. (5.8) and (5.9) to erase $\sum_{i=1}^{N} a_i u_i$, we obtain

$$\lambda = m \left(\frac{\sum_{i=1}^{N} a_i}{N} \pm \sqrt{\frac{\sum_{i=1}^{N} a_i^2}{N}} \right). \tag{5.10}$$

One eigenvalue given by Eq. (5.10) is positive and the other is negative. The other $N-2$ eigenvalues are equal to zero. This is because A^* has rank two. This fact is clear if we rewrite Eq. (5.2) as

$$A^* = \frac{m}{N} \left[\begin{pmatrix} 1 \\ \vdots \\ 1 \end{pmatrix} (a_1 \cdots a_N) + \begin{pmatrix} a_1 \\ \vdots \\ a_N \end{pmatrix} (1 \cdots 1) \right]. \tag{5.11}$$

The largest eigenvalue of A^* is equal to $\lambda = m(\sum_{i=1}^{N} a_i/N + \sqrt{\sum_{i=1}^{N} a_i^2/N})$. This result explains the epidemic threshold that has been derived for the susceptible-infected-susceptible model (Section 6.4) on the activity driven model [Perra *et al.* (2012b); Liu *et al.* (2014)], which coincides with the inverse of the largest eigenvalue of A^*.

The activity driven model has been extended to the case with memory (i.e., correlation between snapshots) [Karsai *et al.* (2014)]. In the extended model, an activated node v_i no longer sends m links to nodes selected randomly with the same probability. Instead, if v_i has been adjacent to k nodes in any of the previous snapshots, each of the m links is connected to one such node with probability $1/(k+c)$, where c is a constant. With the remaining probability $c/(k+c)$, the link connects to one of the $N-k$ nodes randomly selected with the equal probability. This model produces positive correlation between snapshots without affecting $p(k^*)$.

5.4   Priority queue models

Emails continually arrive at our inboxes, and we reply to at least some of them. A single email connects two persons in a temporal way; it is regarded as an event in a temporal network of email users. To understand the bursty nature of event sequences in various temporal network data, it may be helpful to start with analysing behaviour of a single node such as email correspondence patterns of a single user.

If we neglect from whom the email is coming in, a user can be regarded as an information processing system in which time-stamped events arrive, accumulate, and are discharged once a reply, which is also a time-stamped event, is made, as schematically shown in Fig. 5.2. This system is traditionally called queueing process. The output of the queue is a sequence of time-stamped events, corresponding to empirically observed data. Agner Krakup Erlang pioneered theory of queuing processes in the context of telephone traffic congestion early in the twentieth century. In this section, we show that a type of queue model explains the long-tailed behaviour of inter-event times.

The queuing process schematically shown in Fig. 5.2 is not a network. It represents a node, and the unique incoming line is regarded as either a single incoming link or the aggregation of all incoming links. Similar for the unique outgoing line. A set of interconnected queuing processes represents a temporal network (e.g., [Min *et al.* (2009)]). Here we focus on the case of a single queue (Fig. 5.2).

Queuing processes are traditionally formulated in continuous time such that an input, which we call task (circles in Fig. 5.2), arrives according to a Poisson process at a fixed rate, and the task is executed according to a Poisson process at another fixed rate. In queueing theory notation, probably the simplest type of queuing process is denoted by $M/M/1$, where M stands for the Markov process. The first M indicates that the input obeys the Markov process, more specifically, a Poisson process. We denote its rate by λ. The second M indicates that the execution of tasks obeys a Poisson process, whose rate is denoted by μ. Last, 1 in $M/M/1$ indicates that there is just one processing unit. The $M/M/1$ queuing process is a birth-death process, where tasks in the queue are born and die at rates λ and μ, respectively. The master equations for $p_n(t)$, the probability that

Fig. 5.2 Schematic of a queuing process. The square represents the information processing unit. A circle represents a task.

there are n tasks in the queue at time t, are given by

$$\frac{\mathrm{d}p_0(t)}{\mathrm{d}t} = -\lambda p_0(t) + \mu p_1(t), \tag{5.12}$$

$$\frac{\mathrm{d}p_n(t)}{\mathrm{d}t} = -(\lambda + \mu)p_n(t) + \mu p_{n+1}(t) + \lambda p_{n-1}(t). \tag{5.13}$$

The number of tasks, n, performs a continuous-time random walk with a reflecting boundary at $n = 0$. When $\lambda < \mu$, the stationary state exists and is given by

$$\lim_{t \to \infty} p_n(t) = \left(1 - \frac{\lambda}{\mu}\right)\left(\frac{\lambda}{\mu}\right)^n. \tag{5.14}$$

When $\lambda > \mu$ the queue length indefinitely grows as $\approx (\lambda - \mu)t$.

Because empirical inter-event times often obey long-tailed distributions, it may be better to remove the traditional assumption that the inter-arrival time and/or service time obeys the exponential distribution. Then, symbol M is replaced by, for example, D if the time is deterministic (i.e., always the same) or G if the time obeys a general distribution. An $M/G/1$ queuing process implies that the time to the task execution obeys a general (not exponential) distribution. Even in this case, independence of different inter-arrival or service times is assumed, which may be violated in real queuing processes given the fact that inter-event times are often temporally correlated (Section 4.7). In the following, we focus on $M/M/1$ queuing processes unless otherwise stated.

The analysis of queuing processes in the context of power-law distributions of inter-event times mostly treats the waiting time for a task, i.e., time to the execution of the task since it has arrived in the queue. One reason is that the waiting time for a task would also obey a long-tailed distribution [Barabási (2005); Vázquez et al. (2006)]. Another reason is that an $M/M/1$ queuing process produces the exponential distribution of waiting times by definition, regardless of the details of the queue. In the following, we examine waiting times for single tasks. Of course, the waiting time and the inter-event time are different objects, and more data are available for the latter. Discussion to connect the waiting time and inter-event time is found in [Vázquez et al. (2006)].

To discuss the waiting time for a task, we have to specify a rule according to which a task is selected among tasks residing in the queue. Two simple and realistic rules for prioritising task execution are the first-in-first-out (also called first-in-first-served) and random protocols. In the first-in-first-out protocol, the task that has arrived at the queue first is executed first. In the random protocol, each task accumulated in the queue has the same likelihood of being executed. In fact, the waiting time obeys a distribution with an exponential tail for both protocols, contradicting many data involving human behaviour.

On the basis of classical work by Cobham [Cobham (1954)], Barabási showed that the priority queue model, which takes into account properties of human behaviour, produces bursty, long-tailed behaviour in waiting times [Barabási (2005)]. In the model, the processing unit is assumed to have a list of tasks of length L. In each discrete time step, a task always arrives at the queue (therefore different from the $M/M/1$ queueing process). The task is assumed to have a priority x, which is independently drawn from a distribution $\rho(x)$. Then, the task with the largest x value among the L tasks in the list is executed with probability p. With probability $1-p$, a randomly selected task is executed. Because the queue receives one task and executes one task in each time step, the queue length L is preserved over time. A perfect priority queue protocol is produced by $p = 1$, and a random protocol by $p = 0$. Because only the order of x matters for the selection of executed tasks, we assume without loss of generality that $\rho(x)$ is the uniform density on the unit interval $[0, 1]$.

To see mathematically how bursty behaviour emerges from the priority queue model, let us look at the analytically tractable variant of the Barabási's priority queue model. We change the model such that, in each time step, a task with priority x is executed with probability $\Pi(x) \propto x^\gamma$, where $\gamma \geq 0$. The perfect priority queue protocol (i.e., $p \to 1$) is reproduced with $\gamma \to \infty$. The random protocol (i.e., $p = 0$) corresponds to $\gamma = 0$. Let us assume that $\Pi(x)$ is constant in the equilibrium, which is an oversimplification though. Then, the probability that a task with priority x is executed after time τ' since it entered the queue is given by the geometric distribution

$$f(x, \tau') = [1 - \Pi(x)]^{\tau' - 1} \Pi(x). \tag{5.15}$$

Therefore, the average waiting time for a task with priority x, denoted by $\tau(x)$, is given by

$$\tau(x) = \sum_{\tau'=1}^{\infty} \tau' f(x, \tau') = \frac{1}{\Pi(x)} \propto x^{-\gamma}. \tag{5.16}$$

Equation (5.16) is consistent with the intuition that a task with a large priority value x tends to wait for a short period.

Denote by $p^{\mathrm{w}}(\tau)$ the distribution of waiting times, where τ is the waiting time of a task. It should be noted that the waiting time here is that for a task, not that until a next event as in the waiting-time paradox (Section 6.1). For the latter, we will consistently use the notation $\psi^{\mathrm{w}}(\tau)$. We combine Eq. (5.16) with $\rho(x)\mathrm{d}x = p^{\mathrm{w}}(\tau)\mathrm{d}\tau$ to obtain

$$p^{\mathrm{w}}(\tau) \propto \frac{\rho\left(\tau^{-1/\gamma}\right)}{\tau^{1+1/\gamma}}. \tag{5.17}$$

When τ is large, $\rho\left(\tau^{-1/\gamma}\right)$ hardly depends on τ because $\tau^{-1/\gamma}$ is close to zero anyways. Therefore, Eq. (5.17) suggests that $p^{\mathrm{w}}(\tau) \propto \tau^{-1-1/\gamma}$ for large τ. In particular, when the highest priority task is executed almost surely (i.e., $\gamma \to \infty$), we obtain

$$p^{\mathrm{w}}(\tau) \propto \tau^{-1}, \tag{5.18}$$

which is consistent with many real data [Barabási (2005); Vázquez *et al.* (2006)]. When $\gamma = 0$, corresponding to the random protocol, we set $\Pi(x) = 1/L$ to obtain

$$p^{\mathrm{w}}(\tau) = f(x, \tau) = [1 - \Pi(x)]^{\tau-1} \Pi(x) = \left(1 - \frac{1}{L}\right)^{\tau-1} \frac{1}{L}. \tag{5.19}$$

Equation (5.19) indicates that the waiting-time distribution has an exponential tail, consistent with the previous theory. It should be noted that Eq. (5.19) does not depend on x, and therefore, we identified $f(x, \tau)$ with $p^{\mathrm{w}}(\tau)$.

The original Barabási model and its extensions can be studied with the use of diffusion process analysis techniques. Towards this direction, we start with generalising the discrete-time priority queue model to allow stochastic task arrivals and variation in the queue length over time [Grinstein and Linsker (2006)] (also see Supplementary Information of [Barabási (2005)] for numerical results). We now assume that a task with a randomly assigned priority x arrives in the queue with probability λ in each time step. No task arrives with probability $1 - \lambda$. Then, within the same time step, a task in the queue, if any, is selected and executed with probability μ. No task is executed with probability $1 - \mu$ or when the queue is empty. When a task is executed, the task with the largest x value is always selected, corresponding to $p = 1$, i.e., the perfect priority-queue protocol. The following results are known [Grinstein and Linsker (2006, 2008)]:

- If $\lambda = \mu = 1$, each time step involves a task arrival and a task execution such that the queue length L is conserved over time. Regardless of $L(\geq 2)$, we obtain $p^{\mathrm{w}}(\tau) \propto \tau^{-1}$. This result is consistent with Eq. (5.18).
- If $\lambda = \mu < 1$, the queue length L fluctuates over time. In this case, $p^{\mathrm{w}}(\tau) \propto \tau^{-3/2}$.
- If $\mu < \lambda < 1$, tasks accumulate more rapidly than they are executed. The queue length grows linearly in time, i.e., $L \approx (\lambda - \mu)t$. In this case, low-priority tasks with $x < (\lambda - \mu)/\lambda$ are never executed. Tasks with $x \geq (\lambda - \mu)/\lambda$ are executed with the waiting time obeying $p^{\mathrm{w}}(\tau) \propto \tau^{-3/2}$.
- If $\lambda < \mu$, the queue tends to be empty because tasks are executed more rapidly than they accumulate in the queue. In this case, $p^{\mathrm{w}}(\tau) \propto \tau^{-3/2}$ if $\tau \ll \tau_0$ and $p^{\mathrm{w}}(\tau) \propto e^{-\tau/\tau_0}\tau^{-5/2}$ if $\tau \gg \tau_0$, where τ_0 is a characteristic time. In general, τ_0 becomes small as $\mu - \lambda$ increases, such that bursty behaviour is lost.

To obtain these results, the priority queue model is mapped to a biased diffusion model [Grinstein and Linsker (2006)]. Let us glimpse into this idea. We can write

$$p^{\mathrm{w}}(\tau) = \sum_{n=0}^{\infty} \int_0^1 \mathrm{d}x \tilde{Q}(n, x) G(n, x, \tau), \qquad (5.20)$$

where $\tilde{Q}(n, x)$ is the probability that there are n tasks with priority greater than x in the queue in the stationary state (interpretation needs caution when $\lambda > \mu$ or $\lambda = \mu < 1$ in which case the stationary state does not exist), and $G(n, x, \tau)$ is the probability that a task of priority x that arrives in the queue at time $t = 0$ is executed at $t = \tau$, given that there are exactly n tasks at $t = 0$ whose priority value is larger than x. Equation (5.20) is simply the average of the waiting time $G(n, x, \tau)$ over n and x.

To obtain both $\tilde{Q}(n, x)$ and $G(n, x, \tau)$, let $Q(m, x, t)$ be the probability that there are exactly m tasks in the queue whose priority is larger than x at time t. $Q(m, x, t)$ performs a biased random walk in terms of m, with a reflecting boundary at $m = 0$. The master equations are given by

$$Q(m, x, t + 1) = a(x)Q(m + 1, x, t) + b(x)Q(m - 1, x, t)$$
$$+ [1 - a(x) - b(x)] Q(m, x, t), \tag{5.21}$$
$$Q(0, x, t + 1) = a(x)Q(1, x, t) + [1 - a(x)] Q(0, x, t), \tag{5.22}$$

where $a(x)$ and $b(x)$ are the probabilities that m decreases and increases by unity in a time step, respectively. For m to decrease by one, the following two events must happen. First, a task (whose priority is larger than x) must be executed with probability μ. Second, if a task arrives with probability λ, the priority of the new task must not be larger than x. The priority of the new task is larger than x with probability $1 - x$. Therefore,

$$a(x) = \mu \left[1 - \lambda(1 - x) \right]. \tag{5.23}$$

Similarly, we obtain

$$b(x) = \lambda(1 - x)(1 - \mu). \tag{5.24}$$

When $\lambda < \mu < 1$, we combine Eqs. (5.21), (5.22) and $Q(m, x, t + 1) = Q(m, x, t)$ to obtain the stationary density as follows:

$$\tilde{Q}(m, x) = \left[1 - \frac{b(x)}{a(x)} \right] \left[\frac{b(x)}{a(x)} \right]^m, \tag{5.25}$$

which is similar to the result for the $M/M/1$ queuing process (Eq. (5.14)). When $\lambda > \mu$, Eq. (5.25) holds true for high-priority tasks, i.e., those with $x > (\lambda - \mu)/\lambda$. When $\lambda = \mu < 1$, the interpretation of Eq. (5.25) is subtle.

The waiting-time distribution conditioned by n and x, i.e., $G(n, x, \tau)$, is obtained as the first-passage probability of the biased random walk defined by Eqs. (5.21) and (5.22), where the initial condition is given by $Q(m, x, t = 0) = \delta(m - n)$, i.e., there are exactly n tasks with priority larger than x at $t = 0$. The first passage to $m = 0$ determines the waiting time. In this way, the priority queue model is tied to the first-passage time of a one-dimensional biased random walk.

In general, the priority queue model with balanced input and output rates yields the asymptotics $p^{\mathrm{w}}(\tau) \propto \tau^{-1}$ when L is fixed and $p^{\mathrm{w}}(\tau) \propto \tau^{-3/2}$ when L is flexible. Other power-law exponents can be obtained by various extensions of the priority queue model, such as bursty input statistics (i.e., many tasks may arrive in a single time unit) [Masuda *et al.* (2009)]. Some real data obey a waiting-time distribution close to $p^{\mathrm{w}}(\tau) \propto \tau^{-1}$ and others $p^{\mathrm{w}}(\tau) \propto \tau^{-3/2}$. We do not find many data obeying long-tailed $p^{\mathrm{w}}(\tau)$ with other values of the power-law exponent [Vázquez *et al.* (2006)].

5.5   Self-exciting processes

Let us continue to focus on event sequences for a single node. A priority queue model is not the only mechanism that underlies long-tailed behaviour of inter-event times. For example, if an event represents a session of conversation with your friend, it is hard to imagine that you constantly receive conversation tasks to be executed, compare their priority and decide who you should talk to. Rather, you may behave in a self-exciting manner. Once you are engaged in a conversation event, then you may feel like talking with somebody (i.e., the same or other persons) at a higher rate than before. Your inter-event times tend to be short in such an excited state. After some time, you exit a burst of conversation events occurring at the higher rate. Then, you may return to a normal, quiescent state, where the rate of the conversation event is low. In fact, self-exciting mechanisms are capable of producing some properties that empirical inter-event data possess, including long-tailed distributions of inter-event times. In this section, we explain two classes of self-exciting point processes, i.e., the Hawkes processes and two-state cascading Poisson processes.

5.5.1   *Hawkes processes*

The Hawkes processes [Vere-Jones (1970); Hawkes (1971a,b); Hawkes and Oakes (1974)] are a type of non-homogeneous Poisson process in which the instantaneous event rate is modulated by the past occurrence of events. This implies that a Hawkes process is not a renewal process. The event rate at time t, denoted by $\lambda(t)$, is given by

$$\lambda(t) = \lambda_0 + \sum_{\ell, t_\ell \leq t} \phi(t - t_\ell), \qquad (5.26)$$

where t_ℓ is the time of the ℓth event, λ_0 is the basal event rate independent of self excitation, and $\phi(t)$ is the memory kernel describing the additional rate incurred by an event. Causality requires that $\phi(t) = 0$ $(t < 0)$. Hawkes processes have been used for modelling seismological data, online video viewing activities, neural spike trains, genomic data and so forth (see [Masuda *et al.* (2013b)] for references).

The key idea of the Hawkes process is that an event induces an additional event rate. Therefore, $\phi(t)$ should peak at $t = 0$ or small t and decay towards zero as t increases. There are various functional forms for $\phi(t)$ that allow for theoretical analysis and maximum likelihood estimation

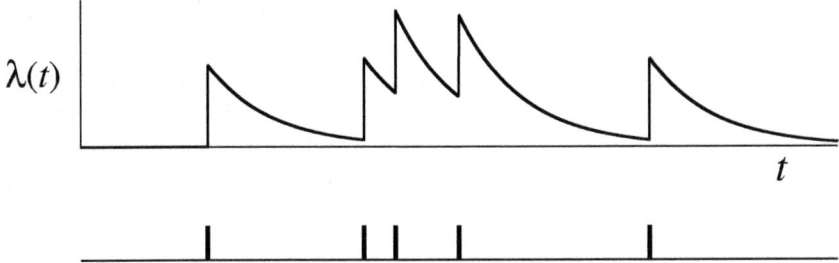

Fig. 5.3 An exemplar time course of event rate $\lambda(t)$ and the corresponding event sequence for a Hawkes process with an exponential memory kernel. In the rastergram shown in the bottom, the event times are shown by ticks.

of the parameter values [Ogata (1999)]. For example, power-law functions such as

$$\phi(t) = \frac{\alpha}{(t - t_{\min})^\theta} \quad (t \geq 0), \tag{5.27}$$

where $\alpha, \theta > 0$, have been used for modelling seismological [Kagan and Knopoff (1981, 1987); Ogata (1988)] and other [Sornette *et al.* (2004)] data. In seismology, Eq. (5.27) is often called Omori law, and the resulting model (in fact, a more labourious version) the epidemic-type aftershock (ETAS) model (for a review, see [Helmstetter and Sornette (2002)]). In this section, we consider a computationally and mathematically simpler, exponential memory kernel given by

$$\phi(t) = \alpha e^{-\beta t} \quad (t \geq 0), \tag{5.28}$$

where $\alpha, \beta > 0$. An exemplar time course of $\lambda(t)$ and the corresponding event sequence are shown in Fig. 5.3.

We define the cluster of events as the set of events that are triggered by a single event that has occurred at the basal rate λ_0. All the events in a cluster are regarded to be descendants of the trigger event, and a cluster corresponds to a burst of events. The expected cluster size is given by [Vere-Jones (1970); Hawkes (1971b)]

$$c = \int_0^\infty \phi(t)dt = \frac{\beta}{\beta - \alpha}. \tag{5.29}$$

The stationary event rate is given by

$$\overline{\lambda} = c\lambda_0 = \frac{\lambda_0 \beta}{\beta - \alpha}. \tag{5.30}$$

The convergence of the event rate requires $\alpha < \beta$. Otherwise, the event rate explodes.

The Hawkes process can also be formulated in the case of static networks [Hawkes (1971a); Pernice *et al.* (2011)]. As in stochastic temporal networks (Section 5.2), we fix the topology of the network for simplicity and focus on modelling of bursty inter-event times on each link. We extend Eq. (5.26) to

$$\boldsymbol{\lambda}(t) = \boldsymbol{\lambda}_0 + \int_0^\infty \boldsymbol{s}(t - t')G(t')\mathrm{d}t'. \tag{5.31}$$

Here, $\boldsymbol{\lambda}(t) = (\lambda_1(t) \cdots \lambda_N(t))$, where $\lambda_i(t)$ is the event rate for the ith node; $\boldsymbol{\lambda}_0 = (\lambda_{0,1} \cdots \lambda_{0,N})$, where $\lambda_{0,i}$ is the basal rate of the process at the ith node. The $N \times N$ matrix $G(t) = (G_{ij}(t))$ is specified by $G_{ij}(t) = A_{ij}\phi_{ij}(t)$, where $A = (A_{ij})$ is the adjacency matrix of the static network, which can be asymmetric, and $\phi_{ij}(t)$ is the memory kernel used to transmit an event that has occurred at the ith node to the jth node. The vector $\boldsymbol{s}(t) = (s_1(t) \cdots s_N(t))$, where $s_i(t) = \sum_{\ell, t_\ell^{(i)} \leq t} \delta(t - t_\ell^{(i)})$ is the event sequence at the ith node, δ is the delta function and $t_\ell^{(i)}$ is the ℓth event generated at the ith node according to rate $\lambda_i(t)$.

At stationarity, Eq. (5.31) is reduced to

$$\overline{\boldsymbol{\lambda}} = \boldsymbol{\lambda}_0 + \int_0^\infty \overline{\boldsymbol{\lambda}}G(t')\mathrm{d}t', \tag{5.32}$$

where $\overline{\boldsymbol{\lambda}} = (\overline{\lambda}_1 \cdots \overline{\lambda}_N)$ and $\overline{\lambda}_i$ is the stationary event rate at the ith node. Equation (5.32) leads to the explicit expression of the stationary event rate:

$$\overline{\boldsymbol{\lambda}} = \boldsymbol{\lambda}_0 \left[I - \int_0^\infty G(t')\mathrm{d}t' \right]^{-1}, \tag{5.33}$$

where I is the identity matrix.

Going back to the case of a single (i.e., non-interacting) process, an advantage of Hawkes processes is that we can derive the maximum likelihood estimation of the parameters for a class of memory kernels. To estimate the parameters of the power-law memory kernel, we have to resort to non-linear numerical optimisation techniques. The estimation for the exponential memory kernel (Eq. (5.28)) is as follows [Ozaki (1979)].

The event rate at time t is written as

$$\lambda(t) = \lambda_0 + \alpha \sum_{\ell=1}^{n} e^{-\beta(t - t_\ell)}, \tag{5.34}$$

where n is the index of the last event before time t. The likelihood of the event sequence $\{t_1, \ldots, t_n\}$ during time span $[0, t_n]$, denoted by $\mathcal{L}(t_1, \ldots, t_n)$, is given by

$$\mathcal{L}(t_1, \ldots, t_n) = \exp\left(-\int_0^{t_n} \lambda(t) dt\right) \prod_{\ell=1}^{n} \lambda(t_\ell). \tag{5.35}$$

By substituting Eq. (5.34) in Eq. (5.35), we obtain the log likelihood as follows:

$$\log \mathcal{L}(t_1, \ldots, t_n) = -\lambda_0 t_n + \sum_{\ell=1}^{n} \frac{\alpha}{\beta} \left[e^{-\beta(t_n - t_\ell)} - 1\right] + \sum_{\ell=1}^{n} \log(\lambda_0 + \alpha A_\ell), \tag{5.36}$$

where

$$A_\ell = \sum_{\ell'=1}^{\ell-1} e^{-\beta(t_\ell - t_{\ell'})}. \tag{5.37}$$

In given data, we often do not know when a point process begins. The process may have started prior to $t = 0$ without yielding events for $t < 0$. Therefore, we assume that the first event is observed at $t_1 = 0$ to slightly modify Eq. (5.36) as

$$\log \mathcal{L}(t_1, \ldots, t_n) = -\lambda_0 t_n + \sum_{\ell=1}^{n} \frac{\alpha}{\beta} \left[e^{-\beta(t_n - t_\ell)} - 1\right] + \sum_{\ell=2}^{n} \log(\lambda_0 + \alpha A_\ell). \tag{5.38}$$

For Eq. (5.38), we obtain

$$\frac{\partial \log \mathcal{L}}{\partial \alpha} = \sum_{\ell=1}^{n} \frac{1}{\beta} \left[e^{-\beta(t_n - t_\ell)} - 1\right] + \sum_{\ell=2}^{n} \frac{A_\ell}{\lambda_0 + \alpha A_\ell}, \tag{5.39}$$

$$\frac{\partial \log \mathcal{L}}{\partial \beta} = -\alpha \sum_{\ell=1}^{n} \left\{ \frac{1}{\beta}(t_n - t_\ell)e^{-\beta(t_n - t_\ell)} + \frac{1}{\beta^2} \left[e^{-\beta(t_n - t_\ell)} - 1\right] \right\}$$

$$- \sum_{\ell=2}^{n} \frac{\alpha B_\ell}{\lambda_0 + \alpha A_\ell}, \tag{5.40}$$

$$\frac{\partial \log \mathcal{L}}{\partial \lambda_0} = -t_n + \sum_{\ell=2}^{n} \frac{1}{\lambda_0 + \alpha A_\ell}, \tag{5.41}$$

where

$$B_\ell = \sum_{\ell'=1}^{\ell-1} (t_\ell - t_{\ell'})e^{-\beta(t_\ell - t_{\ell'})}. \tag{5.42}$$

We obtain the maximum likelihood estimates by setting the left-hand sides of Eqs. (5.39), (5.40) and (5.41) to zero. We carry out the Newton method or the gradient descent method, for example, to estimate α, β and λ_0, starting with some initial guesses.

The Hawkes process with an exponential memory kernel can produce a distribution of inter-event times with large CV values when α ($< \beta$) is close to β. However, the distribution does not look like a power law but a superposition of short-tailed distributions with distinct time scales. In addition, Hawkes processes are poor at capturing the correlation between consecutive inter-event times present in empirical data [Masuda *et al.* (2013b)].

5.5.2   *Cascading Poisson processes*

Let us consider a different class of self-exciting point process models. Malmgren and colleagues proposed two-state point process models to account for long-tailed distributions of inter-event times [Malmgren *et al.* (2008, 2009)].

Their first model, called a cascading non-homogeneous Poisson process, operates as follows [Malmgren *et al.* (2008)]. As in the Hawkes process, in the normal state, an event occurs at a basal rate $\lambda(t) = \lambda_0$. Once an event occurs at rate λ_0, the event rate is raised to $\lambda(t) = \lambda_0 + \Delta\lambda$, where $\Delta\lambda > 0$, representing the excited state. This point process is self-exciting in the sense that an event occurrence in the normal state excites the process to generate a cascade of events at a higher rate. In the excited state, the process is assumed to generate N_a events, where N_a is drawn from a short-tailed distribution (e.g., exponential distribution) $p(N_a)$. After N_a events, the normal state sets in again (Fig. 5.4). In fact, cascading non-homogeneous Poisson processes also assume that $\lambda(t)$ in the normal state is modulated periodically by the circadian and weekly rhythms. Under these assumptions, the model shows long-tailed behaviour of inter-event times.

When applied to data, this model has many parameters to be estimated. For example, the number and positions of the borders between the normal and excited states in terms of the time have to be estimated. Non-homogeneous point process models for neural spike trains often assume two event rates, based on neurophysiological evidence, and are equipped with statistical estimation methods [Chen *et al.* (2009); Tokdar *et al.* (2010); Escola *et al.* (2011)]. Such models may be useful for modelling social and other event sequences as relevant to temporal networks.

A more analytically tractable variant of the model, called a cascading Poisson process, is defined as follows [Malmgren *et al.* (2009)]. The model

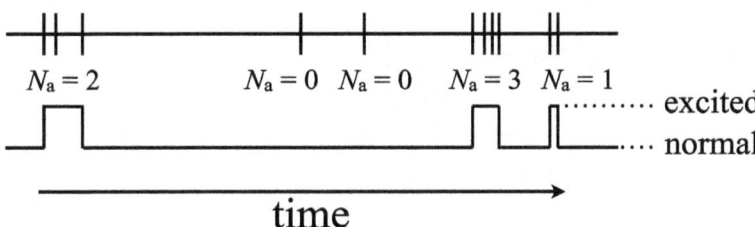

Fig. 5.4   Schematic of the cascading non-homogeneous Poisson process without a periodic rate modulation.

starts with a Poisson process with a basal rate λ_0. Once an event occurs at the basal rate, a further event is generated instantly with probability ξ. The process repeats generating events with probability ξ until it ceases to produce an event with probability $1 - \xi$, thence goes back to the basal Poisson process at rate λ_0. During the additional event generation, the time is not assumed to elapse. In their original paper [Malmgren *et al.* (2009)], the authors estimated λ_0 and ξ by numerically maximising the likelihood for each single-year segment of their empirical data. By concatenating the model across years, they reproduced long-tailed distributions of inter-event times for letter correspondence patterns of celebrities.

Given an event sequence, we can infer the parameters $\theta \equiv \{\lambda_0, \xi\}$ by numerically maximising the likelihood as follows. The distribution of inter-event times is given by

$$
\psi(\tau|\theta) = \begin{cases} \xi & (\tau = 0), \\ (1 - \xi)\lambda_0 e^{-\lambda_0 \tau} & (\tau > 0). \end{cases} \tag{5.43}
$$

Note that multiple events may occur at the same time. To calculate the likelihood, we discretise the time to consider $t_{\max}/t_\star$ time windows, each of whose length is equal to $t_\star$. Then, the main task is to calculate the probability that $N_{t_\star}$ events are produced in a single time window. If there is no event, i.e., $N_{t_\star} = 0$, the process obeys the Poisson process with rate λ_0 such that

$$
p(N_{t_\star} = 0|\theta) = e^{-\lambda_0 t_\star}. \tag{5.44}
$$

Otherwise, $N_{t_\star} - n$ ($0 \leq n \leq N_{t_\star} - 1$) cascades occur in the time window, where a cascade is defined by an event that occurs at a basal rate λ_0 and some (possibly zero) additional events that occur at the same time before the process returns to the basal Poisson process. If $n = 0$, it means that all the cascades did not happen to be self-exciting, incurring no additional

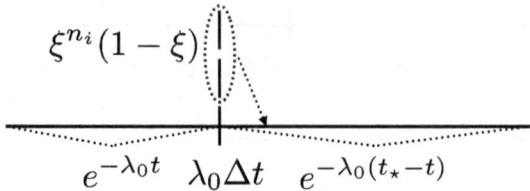

Fig. 5.5   Schematic of a single cascade in a cascading Poisson process. A cascade with $n_i = 2$ occurs at time t.

event. If $n = N_{t_\star} - 1$, there is only one cascade within the time window and all $N_{t_\star}$ events occur at the same time in the single cascade.

Because a cascading Poisson process does not have memory except that multiple events may be produced sequentially at the same time, different cascades are independent of each other. Therefore, we first focus on a single ith cascade ($1 \leq i \leq N_{t_\star} - n$), referring to Fig. 5.5. Assume that the cascade occurs at t ($0 < t < t_\star$) within the time window and that a cascade induces additional n_i events. Then, the cascade produces $n_i + 1$ events as a whole. The probability that no event occurs until time t is equal to $e^{-\lambda_0 t}$. The probability that $n_i + 1$ events are produced and then the process returns to the basal Poisson process in $[t, t + \Delta t]$ is equal to $\lambda_0 \Delta t \xi^{n_i} (1 - \xi)$. The probability that no event occurs between time $t (\approx t + \Delta t)$ and $t_\star$ is equal to $e^{-\lambda_0 (t_\star - t)}$. Therefore, the probability that a single cascade produces $n_i + 1$ events is given by

$$\int_0^{t_\star} e^{-\lambda_0 t} \times \lambda_0 dt \xi^{n_i} (1 - \xi) \times e^{-\lambda_0 (t_\star - t)} = e^{-\lambda_0 t_\star} \lambda_0 t_\star \xi^{n_i} (1 - \xi). \quad (5.45)$$

Because the total number of events, $N_{t_\star}$, is equal to $\sum_{i=1}^{N_{t_\star} - n} (n_i + 1)$, we have to distribute n events, which occur in the self-excited state, to $N_{t_\star} - n$ cascades. Note that some cascades may have no such event. For example, if there are $N_{t_\star} - n = 3$ cascades and $n = 5$ additional events, $(n_1, n_2, n_3) = (2, 1, 2), (1, 2, 2), (0, 4, 1), (5, 0, 0)$ are some of the possible ways to allocate additional events to the three cascades. In fact, there are $\binom{(N_{t_\star} - n) + n - 1}{n} = \binom{N_{t_\star} - 1}{n}$ ways to determine $\{n_i\}$. There are two cautions. First, the $N_{t_\star} - n$ cascades are indistinguishable, resulting in a divisive factor $(N_{t_\star} - n)!$ that should be applied to the likelihood. Second, there are in fact a single basal Poisson process, not $N_{t_\star} - n$ basal Poisson processes running in parallel. By combining these observations with Eq. (5.45), we obtain the probability that we observe $N_{t_\star} - n$ cascades with $N_{t_\star}$ events in total in $[0, t_\star]$ as follows:

$p(N_{t_\star} \text{ events}, N_{t_\star} - n \text{ cascades}|\boldsymbol{\theta})$

$$= \binom{N_{t_\star} - 1}{n} \times \frac{1}{(N_{t_\star} - n)!} \times e^{-\lambda_0 t_\star} \prod_{i=1; \sum_{i=1}^{N_{t_\star}-n} n_i = n}^{N_{t_\star}-n} [\lambda_0 t_\star \xi^{n_i} (1 - \xi)]$$

$$= \binom{N_{t_\star} - 1}{n} e^{-\lambda_0 t_\star} \frac{\xi^n \left[(1 - \xi)\lambda_0 t_\star\right]^{N_{t_\star} - n}}{(N_{t_\star} - n)!}. \tag{5.46}$$

Therefore, we obtain

$$p(N_{t_\star}|\boldsymbol{\theta}) = \sum_{n=0}^{N_{t_\star}-1} p(N_{t_\star} \text{ events}, N_{t_\star} - n \text{ cascades}|\boldsymbol{\theta})$$

$$= e^{-\lambda_0 t_\star} \sum_{n=0}^{N_{t_\star}-1} \binom{N_{t_\star} - 1}{n} \frac{\xi^n \left[(1 - \xi)\lambda_0 t_\star\right]^{N_{t_\star} - n}}{(N_{t_\star} - n)!}. \tag{5.47}$$

A cascading Poisson process in different time windows is independent of each other, reflecting the assumption that events in a time window do not self-excite the process in different time windows. Therefore, we obtain

$$\mathcal{L}(\theta) = \prod_{\ell=0}^{t_{\max}/t_\star - 1} p(N_{t_\star, \ell}|\theta), \tag{5.48}$$

where $\mathcal{L}(\theta)$ denotes the likelihood, and ℓ indexes the time window in the total observation period $t_{\max}$. Because inferring θ based on Eq. (5.48) is analytically infeasible, Eq. (5.48) is numerically maximised given the number of events in each of the $t_{\max}/t_\star$ time windows [Malmgren *et al.* (2009)].

5.6 Markovian log-linear models

The activity driven model in its original form does not incorporate the memory effect; the model assumes that different snapshots are independent of each other. A simple way to incorporate memory into a generative model of temporal networks is to use a Markov process. In the case of the snapshot representation, we make a next snapshot depend on the latest, but not older, snapshot. Social network analysis community in sociology has been concerned with this problem for a long time. Markov models for temporal networks date back at least to seminal work by Holland and Leinhardt [Holland and Leinhardt (1977)]. With the Holland–Leinhardt model included,

continuous-time Markov models have a longer history than discrete-time
ones (e.g., [Snijders (2001)]). However, in this section we focus on discrete-
time Markov models, in particular those represented as logit models. These
models generate temporal networks in the snapshot representation.

For static networks, log-linear models for probabilistic network generation are known under various names including the exponential random
graph models [Wasserman and Pattison (1996); Robins *et al.* (2007); Lusher
et al. (2013)]. Under this framework, the probability that adjacency matrix
A, which can be directed, is generated is given by

$$p(A) = \frac{\sum_{m'=1}^{m} \exp\left[\theta_{m'}\Psi_{m'}(A)\right]}{\sum_{A'} \exp\left[\sum_{m'=1}^{m} \theta_{m'}\Psi_{m'}(A')\right]}, \qquad (5.49)$$

where $\theta_{m'}$ $(1 \leq m' \leq m)$ is a parameter, and $\Psi_{m'}$ $(1 \leq m' \leq m)$ represents a real-valued feature function calculated from a network. The number
of features is denoted by m. The denominator on the right-hand side of
Eq. (5.49) is the normalisation constant, often denoted by Z, and called
the partition function. The summation is taken over all the possible adjacency matrices A', possibly under some constraints that are not explicit
in Eq. (5.49). Equation (5.49) implies that if $\theta_{m'}$ is large, networks whose
m'th feature value, $\Psi_{m'}(A)$, is large is generated with a large probability
and vice versa. In the language of statistical mechanics, Eq. (5.49) is the
canonical ensemble, with $-\theta_{m'}$ playing the role of the inverse temperature
and $\Psi_{m'}(A)$ the energy.

Equation (5.49) has been extended to the case of temporal networks
[Robins and Pattison (2001); Hanneke and Xing (2007); Krackhardt and
Handcock (2007); Hanneke *et al.* (2010)], which can be collectively called
temporal exponential random graph models [Hanneke and Xing (2007);
Hanneke *et al.* (2010)]. The key idea is to combine the exponential random
graph models and the Markov chain to construct manageable first-order
Markovian transition models of networks. Given the adjacency matrix in
the previous time step, $A(t-1)$, the probability of the next adjacency
matrix, $A(t)$, is defined by

$$p(A(t)|A(t-1)) = \frac{\exp\left[\sum_{m'=1}^{m} \theta_{m'}\Psi_{m'}(A(t), A(t-1))\right]}{Z(\boldsymbol{\theta}, A(t-1))}, \qquad (5.50)$$

where the partition function

$$Z(\boldsymbol{\theta}, A(t-1)) = \sum_{A(t)} \exp\left[\sum_{m'=1}^{m} \theta_{m'}\Psi_{m'}(A(t), A(t-1))\right], \qquad (5.51)$$

and $\boldsymbol{\theta} = (\theta_1, \ldots, \theta_m)$. As in the static version (Eq. (5.49)), each $A(t)$ can be an asymmetric matrix.

Examples of the feature functions include the following [Hanneke and Xing (2007); Hanneke *et al.* (2010)]:

- Density: number of directed links present in the snapshot. The feature function is given by

$$\Psi_{\text{density}} = \sum_{i,j=1}^{N} (A(t))_{ij}. \tag{5.52}$$

Because Ψ_{density} does not depend on $A(t-1)$, it is unrelated to memory and can also be used in Eq. (5.49).

- Stability: correlation regarding the presence and absence of each link. The feature function is given by

$$\Psi_{\text{stability}} = \sum_{i,j=1}^{N} \left\{ (A_{ij}(t)A_{ij}(t-1) + [1 - A_{ij}(t)] [1 - (A_{ij}(t-1)] \right\}. \tag{5.53}$$

Note that $\Psi_{\text{stability}}$ is large if consecutive snapshots are similar, i.e., if a link tends to be present (absent) in $A(t)$ when it is present (absent) in $A(t-1)$. If $\theta_{\text{stability}}$ is large, generation of correlated snapshots is encouraged. A negative $\theta_{\text{stability}}$ value emphasises anti-correlated sequences of snapshots. If we assume different coefficients for the two terms inside the summation on the right-hand side of Eq. (5.53), we have the so-called edge-Markovian evolving graph [Clementi *et al.* (2008, 2010)].

- Reciprocity: the tendency that the link from node v_i to node v_j in the previous snapshot results in the reverse link, i.e., link from v_j to v_i in the current snapshot. The feature function is given by

$$\Psi_{\text{reciprocity}} = \frac{\sum_{i,j=1}^{N} A_{ji}(t)A_{ij}(t-1)}{\sum_{i,j=1}^{N} A_{ij}(t-1)}, \tag{5.54}$$

where the denominator on the right-hand side gives a normalisation. If $\theta_{\text{reciprocity}}$ is large, generation of reciprocal links across time is promoted.

- Transitivity: the tendency that link (v_i, v_j) is present when links (v_i, v_ℓ) and (v_j, v_ℓ) are present in the previous snapshot. If such (v_i, v_j) is formed, the triangle composed of nodes v_i, v_j and v_ℓ is closed. This property is common in social networks [Kossinets and Watts (2006)]. The feature function is given by

$$\Psi_{\text{transitivity}} = \frac{\sum_{i,j,\ell=1}^{N} A_{ij}(t) A_{i\ell}(t-1) A_{\ell j}(t-1)}{\sum_{i,j,\ell=1}^{N} A_{i\ell}(t-1) A_{\ell j}(t-1)}, \qquad (5.55)$$

where the denominator on the right-hand side gives a normalisation.

Owing to the Markovian property of the model, the probability of a sequence of snapshots is given by

$$p(\mathcal{A}) = p(A(1)) \prod_{t=2}^{t_{\max}} p(A(t)|A(t-1)), \qquad (5.56)$$

where $\mathcal{A} = \{A(1), \ldots, A(t_{\max})\}$. This expression facilitates numerical simulations of the model. Combined with the log-linear representation of $p(A(t)|A(t-1))$ (Eq. (5.50)), it also enables theoretical analysis and construction of inference and other algorithms for the model.

Both static and temporal exponential random graph models have statistical flavour. In fact, usual maximum likelihood estimation is not tractable because the partition function, Z, involves enumeration of all possible networks and is therefore hard to compute [Robins *et al.* (2007)]. Nevertheless, Markov chain Monte Carlo methods are available to infer the parameters of the model (i.e., $\boldsymbol{\theta}$) given observed or simulated data. By way of likelihood ratios, we can also conduct hypothesis testing, i.e., assessing which model variant is more plausible than others. A typical null model is obtained with a particular $\theta_{m'}$ of interest forced to zero. Then, we can statistically judge if feature $\Psi_{m'}$ is significant or not. In a nutshell, if the likelihood $p(\mathcal{A})$ is considerably increased by keeping the $\theta_{m'}\Psi_{m'}$ term, the m'th feature is important. Otherwise, the $\theta_{m'}\Psi_{m'}$ term should be discarded from statistical points of view. These remarks apply to both static and temporal exponential random graph models.

An important limitation of these classes of models seems to be scalability. As statistical models, inference of parameter values is an indispensable task for these models. If we do that, it seems difficult to treat a large number of nodes under this framework, which contrasts to other generative models that can handle large N.

5.7   Memory networks

Many real temporal networks exhibit correlations between the activation of neighbouring links (Section 4.7). Correlations between link activations imply that certain trajectories, or pathways of information, are favoured while others are discouraged or even forbidden. Such trajectories tend to be poorly reproduced by first-order Markov models. A proper modelling therefore requires a higher-order representation [Rosvall *et al.* (2014); Scholtes *et al.* (2014)]. As we discuss in this section, deviations from a first-order Markov process can significantly alter pathways of diffusion and either accelerate or decelerate diffusion in the network.

To consider this problem, we first map time series of link activation to trajectories of random walkers [Rosvall *et al.* (2014)]. Consider a temporal network in the snapshot representation. We assume that a walker is initially assigned to a randomly selected node and waits there until at least one link is available for hopping. Then, the walker leaves the node with probability $1 - q$ and does not move with probability q, as in the case of the TempoRank centrality measure (Section 4.5.1). If there are multiple possible destinations, the walker selects a link to be traversed with the probability proportional to the link weight. We repeat this process by scanning the snapshots from $t = 1$ to $t = t_{\max}$ many times to generate many trajectories. When $q = 0$, the walker always takes the first available link [Starnini *et al.* (2012); Delvenne *et al.* (2015)]. When $0 < q < 1$, some randomness is introduced in the dynamics [Rocha and Masuda (2014)], which can prevent spurious effects such as a strong tendency of backtracking [Saramäki and Holme (2015)]. As q increases, the ordering of snapshots becomes unimportant and the impact of the temporal network is diluted.

Each generated trajectory is equivalent to a sequence of random variables $\{X_0, X_1, \ldots, X_n\}$, where $X_{n'}$ $(0 \leq n' \leq n)$ is the position of the walker after n' steps. The number of steps, $n(\leq t_{\max})$, depends on an instance of the walk including the choice of the initial node. In general, the probability that the walker visits node v_i after the $(n+1)$th step depends on the full history of the walk process, i.e., $X_0, \ldots, X_n$. In the simple case of the first-order Markov process, denoted by $\mathcal{M}_1$, the transition probability only depends on the current state, X_n (Eq. (2.103)). Under this approximation, the relative weights of links incident to v_i completely determine the walk process, and temporality of the network including correlation between activations of the same or neighbouring links is neglected. The process is described by transition matrix $T = (T_{ij})$ given by

$$T_{ij} = p(i \to j) = \frac{A_{ij}^*}{\sum_{\ell=1}^{N} A_{i\ell}^*}, \tag{5.57}$$

where T_{ij} is the probability that a walker travels from v_i to v_j in one step, and $A^* = \sum_{t=1}^{t_{\max}} A(t)$ is the aggregate adjacency matrix. The probability that the walker visits v_j after $n + 1$ steps is given by

$$p_j(n+1) = \sum_{i=1}^{N} p_i(n)p(i \to j). \tag{5.58}$$

To take into account temporal correlation between link activations, we use the second-order Markov process, denoted by $\mathcal{M}_2$. It is defined on an expanded state space in which each state is a sequence of two consecutively visited nodes, not a node as in $\mathcal{M}_1$ (Fig. 5.6). The $\mathcal{M}_2$ model is identical to a memoryless random walk (i.e., first-order Markov process) between directed links of the original network in which the probability that directed link $\overrightarrow{X_n X_{n+1}}$ is visited depends on $\overrightarrow{X_{n-1} X_n}$, not only on X_n. We regard the state space of the $\mathcal{M}_2$ model as the nodes of a new network, which we call the $\mathcal{M}_2$ or memory network. Regardless of whether the original network is undirected or directed, the memory network has $2M$ nodes and the number of links is proportional to $\langle k^2 \rangle N$ [Evans and Lambiotte (2009)]. A memoryless random walk on the $\mathcal{M}_2$ network corresponds to a random walk with memory on the original network, in which the walker's next location depends on the current and the previous locations.

This procedure can be generalised to higher orders. For example, in a third-order Markov process, the walker's next location depends on the currently visited node v_i and the two previously visited nodes v_j and v_ℓ, and a memory node is specified by $\overrightarrow{\ell j i}$. Here we focus on the second-order Markov process for the following reasons. First, it is conceptually simpler than higher-order Markov processes. Second, gains obtained by the higher-order Markov process are often marginal as compared to their computational cost. Third, for a given amount of data, higher-order Markov processes can be inaccurate because the number of transition probabilities to be estimated exponentially increases with the length of the history considered by the process. With a higher-order Markov process, many transition probabilities are estimated to be zero because there are no instances due to the finite length of data.

Dynamics of a second-order Markov process are encoded by a transition matrix on the $\mathcal{M}_2$ network whose elements are given by $p(\overrightarrow{ij} \to \overrightarrow{jk})$. In practice, $p(\overrightarrow{ij} \to \overrightarrow{jk})$ is estimated by the number of transitions $\overrightarrow{ij} \to \overrightarrow{jk}$ summed over all the generated instances of trajectories, divided by the number of transitions $\overrightarrow{ij} \to \overrightarrow{j\ell}$ summed over all trajectories and ℓ. The transition probability determined in this manner is normalised, i.e., $\sum_{k=1}^{N} p(\overrightarrow{ij} \to \overrightarrow{jk}) = 1$. The transitions are therefore interpreted as movements between links. It should be noted that, even in undirected networks, we must use two mem-

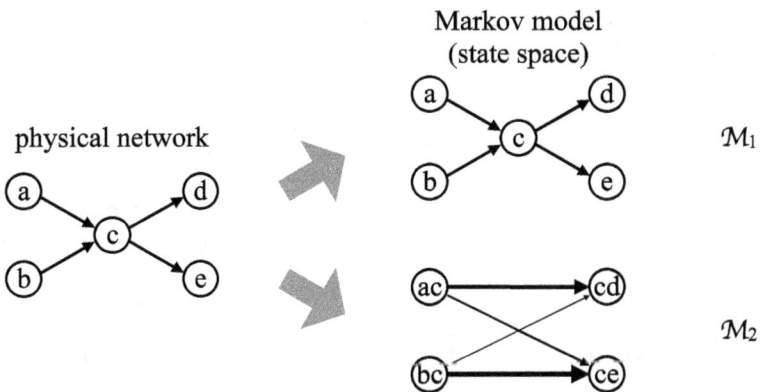

Fig. 5.6   Concept of memory networks. In the first-order Markov model $\mathcal{M}_1$, the state is the node of the physical network, and links indicate transitions between them. In the second-order Markov model $\mathcal{M}_2$, the state is the directed link of the physical network. The state space is structured like the so-called directed line graph of the physical network. If the underlying process is Markovian, transitions in the $\mathcal{M}_2$ network are uniform in the following sense. For example, in the network shown in the figure, node cd would be equally likely to be reached from nodes ac and bc, yielding the same weight for links ac → cd and bc → cd. In the figure, this is not the case, as indicated by the diversified link weights. Therefore, the dynamics in the physical network are non-Markovian.

ory nodes for each pair of connected nodes v_i and v_j because the memory nodes encode the time ordering of the visits.

In the first-order Markov process on an unweighted network, we obtain

$$p(\overrightarrow{ij} \to \overrightarrow{jk}) = \begin{cases} 1/k_j & (v_k \text{ is a neighbour of } v_j), \\ 0 & (\text{otherwise}). \end{cases} \tag{5.59}$$

Equation (5.59) is equivalent to the master equations given by Eq. (5.58). When the dynamics deviate from the first-order Markov process, $p(\overrightarrow{ij} \to \overrightarrow{jk})$ deviates from Eq. (5.59), with certain transitions favoured over other transitions, thereby affecting global trajectories of random walks. In general, the probability that the walker visits node $\overrightarrow{jk}$ after $n + 1$ steps is given by

$$p(\overrightarrow{jk}; n+1) = \sum_i p(\overrightarrow{ij}; n)p(\overrightarrow{ij} \to \overrightarrow{jk}). \tag{5.60}$$

To quantify the second-order Markov constraints on probability flows, we compare the entropy rate of the $\mathcal{M}_1$ and $\mathcal{M}_2$ processes. The entropy rate measures the conditional entropy, i.e., the uncertainty of the next state of the flow given the current state, weighted by the stationary density. For the first-order Markov process, i.e., random walk on the original network, the entropy rate is given by

$$H(X_{n+1}|X_n) = - \sum_{i,j=1}^{N} p_i^* p(i \to j) \log p(i \to j), \qquad (5.61)$$

where $\boldsymbol{p}^* = (p_1^* \ \cdots \ p_N^*)^\top$ is the stationary density. In the second-order Markov process, the entropy rate is the conditional entropy for the memory network and given by

$$H(X_{n+1}|X_n X_{n-1}) = - \sum_{i,j,k=1}^{N} p_{\overrightarrow{ij}}^* p(\overrightarrow{ij} \to \overrightarrow{jk}) \log p(\overrightarrow{ij} \to \overrightarrow{jk}), \qquad (5.62)$$

where $p_{\overrightarrow{ij}}^*$ is the stationary density at node $\overrightarrow{ij}$ in the memory network. In many empirical temporal networks, $H(X_{n+1}|X_n X_{n-1})$ is considerably smaller than $H(X_{n+1}|X_n)$, indicating that the memory is relevant. This observation is consistent with ubiquitous presence of temporal correlations revealed by entropy measures (Section 4.7). The reduction in the entropy with the second-order Markov process implies that the first-order Markov process tends to overestimate the effective number of neighbours in the network.

To summarise, correlations in temporal networks require a modelling beyond the first-order Markov process, restricting pathways of diffusion in the network. This effect of correlation may influence the definition of structural properties of networks as well as dynamics on networks. For example, communities found by second-order pathways based on the Markov stability formalism (Section 3.10.2) tend to be highly synchronised in time. Such communities are undetectable using first-order flows. Temporal correlations also impact transient properties of diffusive processes, in particular, the relaxation time towards a steady state.

The second largest eigenvalue of T in terms of the modulus, denoted by λ_2, both influences the community structure of the network and regulates the relaxation time of the random walk [Delvenne *et al.* (2015)]. The signs of the entries of the corresponding eigenvector, i.e., the Fiedler vector, determine the community to which a node belongs. A perturbation analysis suggests that temporal correlation either increases or decreases λ_2 [Lambiotte *et al.* (2015)]. Suppose that the memory increases λ_2. Then, the random walker obeying the second-order Markov process tends to be more confined in a community of the original network than is suggested by its structure. Such an enhancement in intra-community probability flows

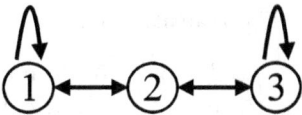

Fig. 5.7   A chain network with $N = 3$ nodes containing self-loops.

pronounces community structure in the corresponding $\mathcal{M}_2$ network. In this case, relaxation is slowed down. If the memory decreases λ_2, the walker transits from community to community more frequently than is suggested by the structure of the original network. Relaxation is accelerated in this situation.

These arguments do not require that the underlying network (i.e., $\mathcal{M}_1$ network) has community structure. For example, consider an undirected chain composed of three nodes and two links denoted by (v_1, v_2) and (v_2, v_3). For a technical reason, we also assume self-loops (v_1, v_1) and (v_3, v_3) (Fig. 5.7).

The random walk without memory (i.e., $\mathcal{M}_1$ process) yields the transition matrix

$$T = \begin{pmatrix} \frac{1}{2} & \frac{1}{2} & 0 \\ \frac{1}{2} & 0 & \frac{1}{2} \\ 0 & \frac{1}{2} & \frac{1}{2} \end{pmatrix}, \tag{5.63}$$

which leads to $\lambda_2 = \pm 1/2$. On this network, consider the $\mathcal{M}_2$ process defined by

$$p(\overrightarrow{11} \to \overrightarrow{11}) = p(\overrightarrow{11} \to \overrightarrow{12}) = p(\overrightarrow{21} \to \overrightarrow{11}) = p(\overrightarrow{21} \to \overrightarrow{12})$$

$$= p(\overrightarrow{33} \to \overrightarrow{33}) = p(\overrightarrow{33} \to \overrightarrow{32}) = p(\overrightarrow{23} \to \overrightarrow{33}) = p(\overrightarrow{23} \to \overrightarrow{32}) = 1/2, \tag{5.64}$$

$$p(\overrightarrow{12} \to \overrightarrow{21}) = p(\overrightarrow{32} \to \overrightarrow{23}) = q, \tag{5.65}$$

$$p(\overrightarrow{12} \to \overrightarrow{23}) = p(\overrightarrow{32} \to \overrightarrow{21}) = 1 - q, \tag{5.66}$$

where $0 < q < 1$. In words, this random walk is memoryless when visiting v_1 and v_3. When the walker visits v_2, it returns to the previous location with probability q and moves to the opposite side of the chain with probability $1 - q$ in the next step. We do not have to distinguish the direction of the link for the two self-loops. Therefore, the $\mathcal{M}_2$ network has six nodes, i.e.,

$\vec{11}$, $\vec{12}$, $\vec{21}$, $\vec{23}$, $\vec{32}$ and $\vec{33}$. The transition matrix is given by

$$T = \begin{pmatrix} \frac{1}{2} & \frac{1}{2} & 0 & 0 & 0 & 0 \\ 0 & 0 & q & 1-q & 0 & 0 \\ \frac{1}{2} & \frac{1}{2} & 0 & 0 & 0 & 0 \\ 0 & 0 & 0 & 0 & \frac{1}{2} & \frac{1}{2} \\ 0 & 0 & 1-q & q & 0 & 0 \\ 0 & 0 & 0 & 0 & \frac{1}{2} & \frac{1}{2} \end{pmatrix}. \tag{5.67}$$

The eigenvalues of T are given by 1, 0, 0, $-1/2$ and $(1 \pm \sqrt{16q-7})/4$. When $q \leq 1/2$, the spectral gap is determined by $\lambda_2 = -1/2$, the same as the case of the $\mathcal{M}_1$ process. When $q > 1/2$, the spectral gap is determined by $\lambda_2 = (1 + \sqrt{16q-7})/4$, which monotonically increases as q increases. A large q value makes λ_2 close to unity and implies that the walker tends to be confined in either of the two communities in the $\mathcal{M}_2$ network, corresponding to either v_1 or v_3 in the $\mathcal{M}_1$ network. It is difficult for the walker to transit from one side of the chain to the other. In particular, $\lim_{q \to 1} \lambda_2 = 1$ such that relaxation occurs infinitely slowly as q approaches unity.

5.8 Metapopulation model

In this section, we introduce the metapopulation model. In short, the model assumes a network of subpopulations, not that of individuals. A subpopulation hosts individuals, and individuals move from one subpopulation to an adjacent subpopulation obeying a rule. In Fig. 5.8, a circle represents an individual, and a box containing individuals represents a subpopulation. The metapopulation framework has been used in ecology [Hanski (1998)] and epidemiology [Anderson and May (1991); Diekmann and Heesterbeek (2000)] for a long time. In network science, metapopulation models have been studied with new analysis tools prone to statistical physics and various mobility data of humans and animals since around 2005 [Hufnagel *et al.* (2004); Colizza *et al.* (2006, 2007)]. In fact, metapopulation models induce temporal interaction between pairs of individuals because, by definition, two individuals meet if and only if they are visiting the same subpopulation. This is why we introduce metapopulation models here. Although metapopulation models are usually considered in conjunction with a dynamical process such as epidemic spreading, in this section we treat them as a network model. An epidemic process model on metapopulation models is explained in Section 6.4.2.

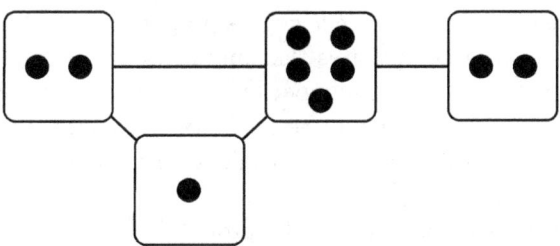

Fig. 5.8   A metapopulation model network with $\tilde{N} = 4$ subpopulations and $N = 10$ individuals.

A metapopulation model network consists of $\tilde{N}$ nodes, called subpopulations, links between pairs of subpopulations, and N particles, which we call individuals. An individual can be anything that is mobile and typically represents a human or animal individual. A subpopulation is a container of individuals, corresponding to a household, office, city, airport, country and so on. The adjacency matrix of the metapopulation model is denoted by $\tilde{A}$ and fixed over time. For simplicity, we assume that it is an undirected network; $\tilde{A}_{ij} = \tilde{A}_{ji}$ $(1 \leq i, j \leq \tilde{N})$. A link between two subpopulations allows flows of individuals between them and may be weighted. A metapopulation model can be regarded as a coarse-grained network of individuals and is practical because detailed connectivity between individuals is often unknown, whereas connectivity between subpopulations may be more accessible. Similarly to networks of individuals, empirical networks of subpopulations typically have heterogeneous degree distributions [Colizza *et al.* (2007)]. Denote by N_i $(1 \leq i \leq \tilde{N})$ the number of individuals in the ith subpopulation, which varies over time t. For any t, $\sum_{i=1}^{\tilde{N}} N_i = N$ is satisfied. A metapopulation model network on $\tilde{N} = 4$ subpopulations and $N = 10$ individuals is shown in Fig. 5.8. We consider continuous time such that mobility of individuals induces a temporal network in the event-based representation on N nodes.

The population is structureless within each subpopulation. An individual interacts with each of the other individuals in the same population at the same rate, regardless of the number of individuals in the subpopulation, or with only a finite number of individuals per unit time [Colizza *et al.* (2007)]. In the former case, the total rate at which an individual in the ith subpopulation interacts with others in the same subpopulation is proportional to $N_i - 1$ because the interaction rate is constant for each of the $N_i - 1$ peers. In the latter case, the total interaction rate does not depend on N_i, and the interaction rate per peer is proportional to $1/(N_i - 1)$. The rational behind this assumption is that an individual may be only capable of interacting with a constant number of others per unit time.

The simplest assumption for the mobility of individuals is to assume that each individual performs an independent continuous-time random walk from subpopulation to subpopulation. In other words, an individual moves to a neighbouring subpopulation with probability $D\Delta t$ in short time Δt. The time to the next movement obeys the independent exponential distribution with mean $1/D$. An individual moves from the ith subpopulation to a neighbouring jth subpopulation with probability $D\Delta t \tilde{A}_{ij}/\tilde{k}_i$, where

$$\tilde{k}_i = \sum_{j=1}^{\tilde{N}} \tilde{A}_{ij} \tag{5.68}$$

is the (weighted) degree of the ith subpopulation.

The number of individuals in each subpopulation, N_i, is approximated to be a continuous variable when N is large. The master equation for N_i is given by

$$\frac{\mathrm{d}N_i}{\mathrm{d}t} = -DN_i + D\sum_{j=1}^{\tilde{N}} N_j \frac{\tilde{A}_{ji}}{\tilde{k}_j}$$

$$= -D\sum_{j=1}^{\tilde{N}} N_j \tilde{L}_{ji}, \tag{5.69}$$

where $1 \leq i \leq \tilde{N}$ and $\tilde{L}$ denotes the random walk normalised Laplacian matrix (Eq. (3.80)) whose elements are given by

$$\tilde{L}_{ij} = \begin{cases} 1 & (i = j), \\ -\frac{\tilde{A}_{ij}}{\tilde{k}_i} & (i \neq j). \end{cases} \tag{5.70}$$

By setting the left-hand side of Eq. (5.69) to zero, we obtain the equilibrium density of individuals as

$$N_i^* = \frac{\tilde{k}_i N}{\langle \tilde{k} \rangle \tilde{N}}. \tag{5.71}$$

Note that $N/\tilde{N}$ is the average population density per subpopulation.

Let us mention some other mobility rules:

- The diffusion rate may depend on the degrees of the two adjacent subpopulations [Colizza and Vespignani (2008)].

- The diffusion rate may be determined by the number of particles in the source subpopulation, N_i, called the zero-range process [Evans (2000); Colizza and Vespignani (2008)]. In particular, if the diffusion constant is given by DN_i^γ for the ith subpopulation, the master equation is altered to

$$\frac{\mathrm{d}N_i}{\mathrm{d}t} = -(DN_i^\gamma)N_i + \sum_{j=1}^{\tilde{N}} \frac{\tilde{A}_{ji}}{\tilde{k}_j}(DN_j^\gamma)N_j$$

$$= -DN_i^{\gamma+1} + D\sum_{j=1}^{\tilde{N}} \frac{\tilde{A}_{ji}}{\tilde{k}_j}N_j^{\gamma+1}. \qquad (5.72)$$

When $\gamma = 0$, individuals are non-interacting in deciding on the movement, and Eq. (5.72) reduces to Eq. (5.69). When $\gamma > 0$, interaction is repulsive such that individuals would move away from subpopulations with a high population density. When $\gamma < 0$, individuals are attracted to highly populated subpopulations. When $\gamma = -1$, the right-hand side of Eq. (5.72) is always equal to zero such that any initial condition is a stationary state [Colizza and Vespignani (2008)]. When γ is small, condensation of individuals, in which a subpopulation attracts $O(N)$ individuals, can occur [Evans (2000); Noh *et al.* (2005)].

- Human behaviour generally induces recurrent mobility patterns back and forth between home and workplaces/schools [Balcan and Vespignani (2011); Belik *et al.* (2011)]. Suppose that a home subpopulation is chosen by each individual with the stationary density given by Eq. (5.71) and then individuals commute to an adjacent subpopulation and come back to their home subpopulations [Balcan and Vespignani (2011)]. We can justify the model by considering that the home subpopulation is selected according to the simple diffusion on a very slow time scale and that individuals travel back and forth between the home subpopulation and the adjacent subpopulation on a fast time scale. Differently from the previous rules, the individuals memorise their own home subpopulations throughout the dynamics.

We assume that an individual homed at the ith subpopulation diffuses to an adjacent jth subpopulation at rate σ_{ij} and that an individual visiting the jth subpopulation returns to the home subpopulation at rate $1/\tau_i$. Denote by N_{ij} the number of individuals that are homed at the ith subpopulation visiting an adjacent jth subpopulation. Therefore, N_{ii} represents the number individuals located at their home subpopulation, which is the ith subpopulation. The meaning of the variables is

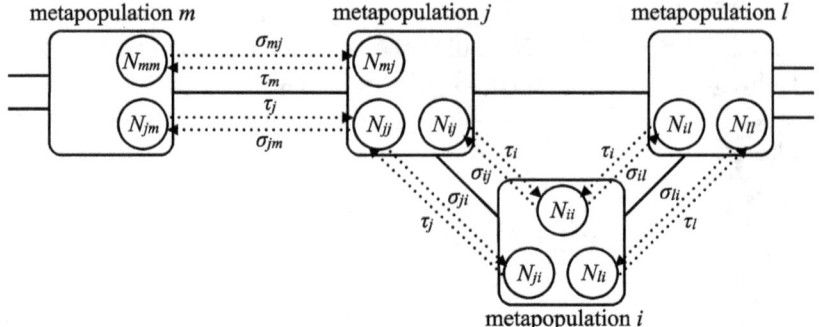

Fig. 5.9   Schematic of a metapopulation model with recurrent mobility patterns.

schematically shown in Fig. 5.9. The master equations are given by

$$\frac{\mathrm{d}N_{ii}}{\mathrm{d}t} = -\,\sigma_i N_{ii} + \frac{1}{\tau_i} \sum_{j=1,j\neq i}^{\tilde{N}} \tilde{A}_{ij} N_{ij}, \tag{5.73}$$

$$\frac{\mathrm{d}N_{ij}}{\mathrm{d}t} = \sigma_{ij} N_{ii} - \frac{1}{\tau_i} N_{ij} \quad (j \neq i, \tilde{A}_{ij} = 1), \tag{5.74}$$

where $1 \leq i, j \leq \tilde{N}$ and $\sigma_i = \sum_{j=1;j\neq i}^{\tilde{N}} \tilde{A}_{ij}\sigma_{ij}$.
By setting the right-hand sides of Eqs. (5.73) and (5.74) to zero and using $N_i = N_{ii} + \sum_{j=1,j\neq i}^{\tilde{N}} \tilde{A}_{ij} N_{ij}$ and Eq. (5.71), we obtain the stationary density as follows:

$$N_{ii}^* = \frac{\tilde{k}N}{\langle k \rangle \tilde{N}(1 + \sigma_i \tau_i)}, \tag{5.75}$$

$$N_{ij}^* = \frac{\tilde{k}N\sigma_{ij}\tau_i}{\langle k \rangle \tilde{N}(1 + \sigma_i \tau_i)}. \tag{5.76}$$

The stationary number of individuals visiting the jth subpopulation regardless of the home subpopulation (i.e., N_j^*) is given by $N_{jj}^* + \sum_{i=1;i\neq j}^{\tilde{N}} N_{ij}^*$.

Chapter 6

Dynamics on temporal networks

In this chapter, we review representative dynamical processes taking place on temporal networks.

For static networks, dynamics of networks and dynamics on networks are relatively distinct processes. We described dynamics of networks in detail in Chapters 4 and 5. Growing network models and adaptive network models, which we do not seriously consider in this book, are also examples of dynamics of networks. Dynamics on (static) networks are a dynamical process occurring on static networks, such as random walk, epidemic processes and synchronisation dynamics.

For temporal networks, the distinction is less clear. Dynamics of networks are given by temporal network data or models. When we talk about dynamics on temporal networks, we need to be careful because events in temporal networks are often recorded as a result of some dynamics. For example, if an event is defined as an email sent from one person to another, the event also carries information. Then, a sequence of events naturally defines viral spreading as well as a temporal network. Similarly, an event may be marked as the motion of a random walker from one node to another. This type of ambivalent interpretation is relevant especially when we consider the event-based representation of temporal networks.

Another aspect of dynamics on temporal networks is the relative time scale of dynamical processes and dynamics of networks. Static networks are the limit of infinitely slow dynamics of networks, or in the quenched regime [Porter and Gleeson (2014)]. Then, the time scale of dynamical processes can be arbitrarily changed without loss of generality. For example, in epidemic process models (Section 6.4.1), it is usual to set the recovery rate to unity because multiplying the infection and recovery rates by the same factor corresponds to a rescaling of the time.

However, such an operation is not allowed for dynamics on temporal networks. Multiplying the infection and recovery rates by the common value larger than unity makes the epidemic dynamics relatively faster than the dynamics of networks as compared to before the multiplication. A particular phenomenon observed for a dynamical process on a temporal network may disappear if the relative time scale of the two dynamics is changed.

In addition, if the dynamics of networks are much faster than dynamics on networks, we may be able to reliably observe mean quantities on networks, such as the average degree of each node, but not full dynamics of the detailed network structure. Then, the best representation of the temporal network may be to employ a random graph model (e.g., configuration model), corresponding to the annealed regime [Porter and Gleeson (2014)].

In this chapter, we consider dynamical processes on temporal networks. Because events are often recorded as a result of dynamics, as mentioned above, we sometimes assume that an event always induces a step of the dynamical process under question (e.g., movement of a random walker).

6.1 Waiting-time paradox

Many interesting results regarding dynamics on temporal networks are consequences of the waiting-time paradox in queuing theory [Feller (1971); Allen (1990)]. This phenomenon is a purely mathematical effect and is relevant to the event-based representation of temporal networks. The paradox dictates that if a dynamical object (e.g., random walker, wave of epidemic, interface between different opinions) arrives at a node from some other node, the object is forced to wait typically for a long time before the next event. Despite that an arrival at a node occurs uniformly within an inter-event time, the average waiting time is longer than a half of the average inter-event time. This is true unless the inter-event time is constant, which implies a periodic event sequence, a very special case. The paradox also occurs for the exponential distribution of inter-event times, i.e., Poisson processes. The waiting-time paradox is not a network phenomenon. Anecdotally, the average waiting time for a passenger of a bus since the passenger has arrived at the bus stop is longer than half of the average interval between two buses expected from the timetable. This is why the waiting-time paradox is also called the bus paradox.

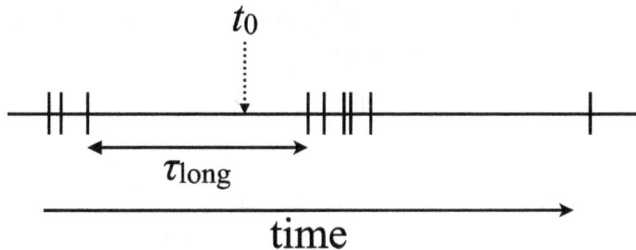

Fig. 6.1   Waiting-time paradox. A randomly drawn time tends to fall within a long interval.

The difference between the average waiting time and the average inter-event time is large if the inter-event time obeys a long-tailed distribution. This is mathematically the same as the friendship paradox (Section 3.2). In the friendship paradox, a node is sampled with the probability proportional to its size, i.e., degree. In the waiting-time paradox, a time within two events is sampled with the probability proportional to the size of the interval, i.e., inter-event time.

Suppose that the inter-event time, τ, is independently distributed for all links according to probability density $\psi(\tau)$. For Poisson processes, $\psi(\tau)$ is the exponential distribution. Consider a random walker arriving at node v_1 from somewhere else. If the next event can only occur on link (v_1, v_2), what is the waiting time before the walker transits to v_2? Of course, the answer is a random variable. The time at which the walker arrives at v_1, denoted by t_0, is uniformly distributed on the time axis if no additional information is given. Figure 6.1 schematically represents an event sequence on link (v_1, v_2). As this figure shows, t_0 is located inside an interval with a probability proportional to the length of the interval. If t_0 falls in a relatively long interval whose inter-event time is τ_{long}, the conditional mean waiting time is equal to $\tau_{\text{long}}/2$, which is large. Conversely, if t_0 falls in a short interval with inter-event time τ_{short}, the conditional mean waiting time, equal to $\tau_{\text{short}}/2$, is short. Crucially, the former event is more likely to occur than the latter event with a factor of $\tau_{\text{long}}/\tau_{\text{short}}$. Therefore, we expect that the mean waiting time is longer than $\langle \tau \rangle/2$.

The distribution of waiting times, $\psi^{\text{w}}(t)$, is derived as follows. The probability density with which t_0 falls in an interval of length τ is given by

$$f(\tau) = \frac{\tau\psi(\tau)}{\int_0^\infty \tau'\psi(\tau')\mathrm{d}\tau'} = \frac{\tau\psi(\tau)}{\langle \tau \rangle}. \tag{6.1}$$

Under the condition that this event happens, the waiting time is equal to t with the probability density given by

$$g(t|\tau) = \begin{cases} 1/\tau & (0 \le t \le \tau), \\ 0 & (\tau > t). \end{cases} \tag{6.2}$$

By combining Eqs. (6.1) and (6.2), we obtain

$$\psi^{\mathrm{w}}(t) = \int_t^\infty f(\tau)g(t|\tau)\mathrm{d}\tau = \frac{1}{\langle \tau \rangle} \int_t^\infty \psi(\tau)\mathrm{d}\tau. \tag{6.3}$$

The mean waiting time is given by

$$\int_0^\infty t\psi^{\mathrm{w}}(t)\mathrm{d}t = \frac{1}{\langle \tau \rangle} \int_0^\infty \left[t \int_t^\infty \psi(\tau)\mathrm{d}\tau \right] \mathrm{d}t$$

$$= \frac{1}{\langle \tau \rangle} \int_0^\infty \psi(\tau) \left[\int_0^\tau t\mathrm{d}t \right] \mathrm{d}\tau$$

$$= \frac{\langle \tau^2 \rangle}{2\langle \tau \rangle}. \tag{6.4}$$

When the event sequence is perfectly periodic such that $\psi(\tau)$ is the delta function, we obtain $\langle \tau^2 \rangle = \langle \tau \rangle^2$ and the mean waiting time is equal to $\langle \tau \rangle/2$. No paradox occurs in this case. However, even in the case of Poisson processes, i.e., $\psi(\tau) = \lambda e^{-\lambda t}$, we obtain $\langle \tau \rangle = 1/\lambda$ and $\langle \tau^2 \rangle = 2/\lambda^2$ such that the mean waiting time is equal to $1/\lambda = \langle \tau \rangle$, i.e., twice the case of the periodic sequence with the same mean inter-event time. For long-tailed $\psi(\tau)$, $\langle \tau^2 \rangle$ is much larger than $\langle \tau \rangle$. Therefore, the waiting time is typically very long. In particular, if $\langle \tau^2 \rangle$ diverges, the mean waiting time also diverges. It is the case when $\psi(\tau) \propto \tau^{-\gamma}$ with $\gamma \le 3$. If $2 < \gamma \le 3$, the mean inter-event time is finite, while the mean waiting time is infinite.

What is the implication of the waiting-time paradox in temporal networks? It mostly concerns dynamics on temporal networks because a waiting time is drawn when something (i.e., a step in a dynamical process) happens. Consider a random walker that has moved from v_i to v_j at time t. It may traverse to a different node v_ℓ ($\ell \neq i$) through link (v_j, v_ℓ) if the link has an event within a short time. At time t, the waiting time on link (v_j, v_ℓ) should be drawn from $\psi^{\mathrm{w}}(\tau)$, not from $\psi(\tau)$ (Section 6.3). This is because the last event on link (v_j, v_ℓ) occurred sometime before t, not exactly at t. Similar reasoning applies to other dynamics on temporal networks. Because the waiting time is much longer than the inter-event time in various data, various dynamical processes running on temporal networks would proceed more slowly on temporal than the corresponding aggregate networks. For example, in a majority of cases, epidemic waves take longer time to spread [Holme and Saramäki (2012); Masuda and Holme (2013)]. The same for the voter model of opinion formation [Fernández-Gracia *et al.* (2011); Takaguchi and Masuda (2011)]. We remark that the waiting-time paradox is not a necessary condition for dynamics to slow down. Slowing down due to temporal networks can occur even under the snapshot representation in which different snapshots are uncorrelated (Section 6.5).

The waiting-time paradox also warns us of the choice of the initial condition when we run renewal processes [Cox (1962)]. If we draw the first inter-event time at $t = 0$ from $\psi(\tau)$, a renewal process is called the ordinary renewal process. An ordinary renewal process assumes that the event has occurred at $t = 0$. This assumption is probably unnatural when we run dynamics on temporal networks, where many renewal processes are running in parallel and each renewal process produces a series of events between a pair of nodes. In reality, all pairs of individuals would not be simultaneously involved in a conversation event at $t = 0$. An alternative model is the equilibrium renewal process, in which the time to the first event is drawn from $\psi^{\mathrm{w}}(\tau)$ at $t = 0$. The second and later inter-event times are drawn from $\psi(\tau)$. An equilibrium renewal process assumes that the process started at $t = -\infty$ and therefore the equilibrium has been reached at $t = 0$. In other words, the information about the first event at $t = -\infty$ has been forgotten when the dynamical process of interest starts at $t = 0$.

6.2   Gillespie algorithms

When we numerically simulate a dynamical process on temporal networks in the event-based representation, we have to know on which node/link and when the next event in the entire network takes place. For example, node v_1 may contract infection from an infected neighbour v_2, while v_2 may recover. Which event occurs first?

The Gillespie algorithm, originally designed for simulating systems of chemical reactions, is a practical exact algorithm for Markovian dynamics, i.e., the case in which the occurrence of each candidate event obeys a Poisson process at generally a different rate [Kendall (1950); Gillespie (1976, 1977)]. Nonetheless, empirical temporal network data usually deviate from the Poissonian statistics. In this section, we present the Gillespie algorithm and its extensions to the case of general distributions of inter-event times.

To describe the ordinary Gillespie algorithm, consider a set of N independent Poisson processes with rate λ_i ($1 \leq i \leq N$). Each Poisson process is attached to a node or link and generates event sequences. The Gillespie algorithm generates a sequence of time τ to the next event in any of the N Poisson processes and also informs which of the N processes actually generates the next event after τ (Fig. 6.2). Because of the independence of different Poisson processes, superposition of the N Poisson processes is a single Poisson process with rate $N\langle\lambda\rangle \equiv \sum_{i=1}^{N} \lambda_i$. Therefore, τ is drawn from the exponential distribution given by

$$\phi(\tau) = N\langle\lambda\rangle e^{-N\langle\lambda\rangle\tau}. \tag{6.5}$$

With this τ, the ith process produces the next event with probability

$$\Pi_i = \frac{\lambda_i}{N\langle\lambda\rangle}. \tag{6.6}$$

By drawing τ and i according to the distributions given by Eqs. (6.5) and (6.6), respectively, we generate one event. To draw τ, we note that the survival probability of $\phi(\tau)$ is given by $\int_{\tau}^{\infty} \phi(\tau')d\tau' = e^{-N\langle\lambda\rangle\tau}$. Therefore, we draw a random variable u according to the uniform density on $[0, 1]$ and determine τ according to $u = e^{-N\langle\lambda\rangle\tau}$, i.e., $\tau = -\log u/N\langle\lambda\rangle$.

Then, we advance the time by τ. If this event has changed the event rate for some processes j, where j may be different from i, we accordingly modify λ_j. For example, if a susceptible (i.e., healthy) node v used to be adjacent to two infected nodes and one of the two infected nodes has recovered upon an event, the rate at which v gets infected usually decreases. Then, we have to reduce the corresponding λ_j although v has not undergone any event itself. Next, we draw the next τ and i according to Eqs. (6.5) and (6.6) again. This procedure is justified because of the memoryless property of Poisson processes. In other words, the time to the next event for each process does not depend on what has happened in the past.

The non-Markovian Gillespie algorithm is an extension of the Gillespie algorithm to the case of general distributions of inter-event times [Boguñá *et al.* (2014)]. We denote by $\psi_i(\tau)$ the distribution of inter-event times for the ith renewal process ($1 \leq i \leq N$). If $\psi_i(\tau)$ is the exponential distribution, we recover a Poisson process. To generate event sequences statistically correctly, we must know the time since the last event for each process, which is denoted by t_i. The probability density for the waiting time for the ith process, when there is no other process running in parallel, is given by

$$\psi_i^{\mathrm{w}}(\tau|t_i) = \frac{\psi_i(t_i + \tau)}{\Psi_i(t_i)}, \tag{6.7}$$

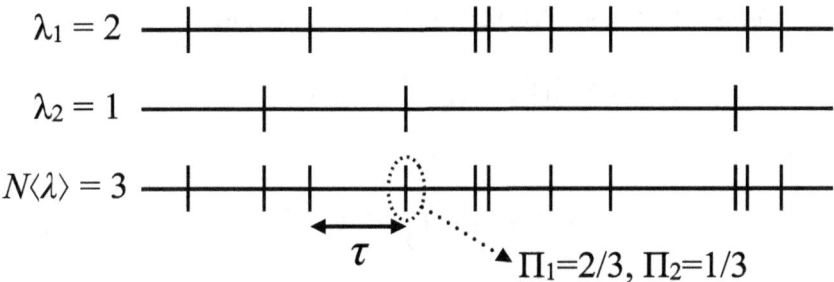

τ

$\Pi_1 = 2/3,\ \Pi_2 = 1/3$

Fig. 6.2 Schematic of the Gillespie algorithm. $N = 2$ Poisson processes, one with $\lambda_1 = 2$ and the other with $\lambda_2 = 1$, are running in parallel. The superposed Poisson process has rate 3 and determines the next event (in the dotted circle). The event is judged to originate from process 1 with probability $\Pi_1 = \lambda_1/N\langle\lambda\rangle = 2/3$ and process 2 with probability $\Pi_2 = \lambda_2/N\langle\lambda\rangle = 1/3$.

where

$$\Psi_i(t_i) = \int_{t_i}^{\infty} \psi_i(\tau')\mathrm{d}\tau' \qquad (6.8)$$

is the survival probability, i.e., the probability that the inter-event time is larger than t_i.

Because the ith process coexists with the other $N-1$ processes, we want to compute the probability that the next event in the entire population of the N processes occurs in the ith process at a certain time τ. The distribution of τ is not given by Eq. (6.7) because one of the other processes may generate an event prior to the ith process does. We denote by $\phi(\tau, i|\{t_j\})$ the probability density with which the ith process, not any other process, generates the next event with waiting time τ under the condition that the last event generated by the jth process ($1 \leq j \leq N$) has occurred exactly time t_j ago. We obtain

$$\phi(\tau, i|\{t_j\}) = \psi_i^{\mathrm{w}}(\tau|t_i) \prod_{j=1; j \neq i}^{N} \Psi_j(\tau|t_j), \qquad (6.9)$$

where

$$\Psi_j(\tau|t_j) = \int_{\tau}^{\infty} \psi_j^{\mathrm{w}}(\tau'|t_j)\mathrm{d}\tau' = \frac{\Psi_j(t_j + \tau)}{\Psi_j(t_j)} \qquad (6.10)$$

is the conditional survival probability, i.e., the probability that the waiting time of the jth process running in isolation is larger than τ, given that the last event occurred time t_j ago. Equation (6.9) states that the ith process generates an event within a small time window around τ, whereas the other $N-1$ processes do not generate an event until this time.

Using Eqs. (6.7) and (6.10), we rewrite Eq. (6.9) as

$$\phi(\tau, i | \{t_j\}) = \frac{\psi_i(t_i + \tau)}{\Psi_i(t_i + \tau)} \Phi(\tau | \{t_j\}), \qquad (6.11)$$

where

$$\Phi(\tau | \{t_j\}) = \prod_{j=1}^{N} \frac{\Psi_j(t_j + \tau)}{\Psi_j(t_j)}. \qquad (6.12)$$

We can interpret Eq. (6.11) as follows. First, Eq. (6.12) is equal to the survival probability for the set of the N processes. In other words, the probability that no process generates an event for next time τ is equal to Eq. (6.12). Therefore, this expression can be used to draw a next event time as we do for the original Gillespie algorithm using Eq. (6.5). Second, once the next event time τ is determined, the prefactor $\psi_i(t_i + \tau)/\Psi_i(t_i + \tau)$ on the right-hand side of Eq. (6.11) is proportional to the probability that the ith process generates that event. This fact is also verified by

$$\Pi_i \equiv \frac{\phi(\tau, i | \{t_j\})}{\sum_{j=1}^{N} \phi(\tau, j | \{t_j\})} = \frac{\lambda_i(t_i + \tau)}{\sum_{j=1}^{N} \lambda_j(t_j + \tau)}, \qquad (6.13)$$

where

$$\lambda_i(t) = \frac{\psi_i(t)}{\Psi_i(t)} \qquad (6.14)$$

is the instantaneous rate of the ith process.

Collection of these results defines an extension of the Gillespie algorithm carried out in the following steps. For simplicity, we assume ordinary renewal processes, in which all processes generate an event at $t = 0$ (Section 6.1).

(1) Initialise $t_j = 0$ $(1 \leq j \leq N)$.
(2) Draw the time to the next event from the cumulative distribution $\Phi(\tau | \{t_j\})$. Concretely, generate a random variable u from the uniform density on the unit interval and equate it to $\Phi(\tau | \{t_j\})$ to determine τ.
(3) Determine the process, indexed by i, which generates the next event using Eq. (6.13).
(4) Update the list of the time since the last event as $t_j \rightarrow t_j + \tau$ $(j \neq i)$ and $t_i = 0$. If there are processes j $(1 \leq j \leq N)$ whose distribution of inter-event times has changed upon the event, update $\psi_j(\tau)$ and reset $t_j = 0$ if necessary.

The non-Markovian Gillespie algorithm approximates the aforementioned extension of the Gillespie algorithm and is computationally feasible for large N and exact in the limit $N \to \infty$. To derive the non-Markovian Gillespie algorithm, we rewrite Eq. (6.12) as

$$\Phi(\tau|\{t_j\}) = \exp\left[-\sum_{j=1}^{N} \ln \frac{\Psi_j(t_j)}{\Psi_j(t_j + \tau)}\right]. \tag{6.15}$$

When N is large, it is unlikely that all processes wait for more than short τ before generating an event. Therefore, a realised τ value would be very small, and $\Phi(\tau|\{t_j\})$ is tiny except for $\tau \approx 0$. Using this fact, we expand Eq. (6.15) around $\tau = 0$. By using $\Psi_j(t_j + \tau) = \Psi_j(t_j) - \psi_j(t_j)\tau + O(\tau^2)$ $(1 \leq j \leq N)$, which results from differentiation of Eq. (6.8), we obtain

$$\Phi(\tau|\{t_j\}) = \exp\left[-\sum_{j=1}^{N} \ln \frac{\Psi_j(t_j)}{\Psi_j(t_j) - \psi_j(t_j)\tau + O(\tau^2)}\right]$$

$$= \exp\left\{-\sum_{j=1}^{N} \ln\left[1 + \frac{\psi_j(t_j)}{\Psi_j(t_j)}\tau + O(\tau^2)\right]\right\}$$

$$= \exp\left[-\sum_{j=1}^{N} \frac{\psi_j(t_j)}{\Psi_j(t_j)}\tau + O(\tau^2)\right]$$

$$\approx \exp\left[-\tau N\overline{\lambda}(\{t_j\})\right], \tag{6.16}$$

where the average instantaneous rate is given by

$$\overline{\lambda}(\{t_j\}) = \frac{\sum_{j=1}^{N} \lambda_j(t_j)}{N} = \frac{1}{N} \sum_{j=1}^{N} \frac{\psi_j(t_j)}{\Psi_j(t_j)}. \tag{6.17}$$

For a technical reason, we have to remove the process that has generated the last event from the summation in Eq. (6.17).

With this approximation, the time to the next event is determined by $\Phi(\tau|\{t_j\}) \approx \exp\left[-\tau N\overline{\lambda}(\{t_j\})\right] = u$, i.e., $\tau = -\ln u/N\overline{\lambda}(\{t_j\})$. The process that generates this event is determined by setting $\tau = 0$ in Eq.(6.13), i.e.,

$$\Pi_i = \frac{\lambda_i(t_i)}{N\overline{\lambda}(\{t_j\})}. \tag{6.18}$$

These relationships define the non-Markovian Gillespie algorithm [Boguñá *et al.* (2014)]. For the Poisson processes, we set $\lambda_i(t_i) = \lambda_i$ to recover the original Gillespie algorithm given by Eqs. (6.5) and (6.6).

In the non-Markovian Gillespie algorithm, we have to update the instantaneous event rates of all processes as each event occurs, which is time consuming. An alternative generalisation of the Gillespie algorithm can be derived on the basis of the Laplace transform [Masuda and Rocha (2016)]. To explain the algorithm, which we call the Laplace Gillespie algorithm, we first consider a single renewal process. When an event occurs, we first draw the rate of the Poisson process, denoted by λ, according to a fixed probability density function $p(\lambda)$. Then, we draw the time of the next event according to the Poisson process with rate λ. Upon the occurrence of the next event, we renew the rate λ by redrawing it from $p(\lambda)$. We repeat these procedures.

For an ensemble of N renewal processes, we denote the density of the event rate for the ith process by $p_i(\lambda_i)$. The Laplace Gillespie algorithm proceeds as follows:

(1) Initialise each of the N processes by drawing the rate λ_i ($1 \leq i \leq N$) of the ith Poisson process from $p_i(\lambda_i)$.
(2) Determine the time to the next event by $\tau = -\ln u / \sum_{j=1}^{N} \lambda_j$, where u is uniformly distributed on $[0,1]$.
(3) Select the process that generates the next the event with probability $\Pi_i = \lambda_i / \sum_{j=1}^{N} \lambda_j$.
(4) Draw a new rate λ_i according to probability density $p_i(\lambda_i)$. If there are processes j ($1 \leq j \leq N$) whose distribution of inter-event times has changed upon the event, update $p_j(\lambda_j)$ and redraw λ_j from the new $p_j(\lambda_j)$. The rates of the other processes are unchanged.
(5) Repeat steps (2)–(4).

For a given density of the event rate $p(\lambda)$, the process generated by this algorithm is a renewal process. The probability density of inter-event times, $\psi(\tau)$, is given by

$$\psi(\tau) = \int_0^\infty p(\lambda) \lambda e^{-\lambda \tau} \mathrm{d}\lambda. \qquad (6.19)$$

Integration of both sides of Eq. (6.19) yields

$$\Psi(\tau) = \int_\tau^\infty \psi(\tau') \mathrm{d}\tau' = \int_0^\infty p(\lambda) e^{-\lambda \tau} \mathrm{d}\lambda. \qquad (6.20)$$

Equation (6.20) indicates that the survival probability of inter-event times, $\Psi(\tau)$, is the Laplace transform of $p(\lambda)$. Therefore, a renewal process can be simulated by the Laplace Gillespie algorithm if $\Psi(\tau)$ is the Laplace transform of a probability density function on non-negative values.

A necessary and sufficient condition for the existence of $p(\lambda)$ is that $\Psi(\tau)$ is completely monotone [Feller (1971)] and that $\Psi(0) = 1$, where the complete monotonicity is defined by

$$(-1)^n \frac{d^n \Psi(\tau)}{d\tau^n} \geq 0 \quad (\tau \geq 0, n = 0, 1, \ldots). \tag{6.21}$$

Condition $\Psi(0) = 1$ is always satisfied. Equations (6.21) with $n = 0$ and $n = 1$ read $\Psi(\tau) \geq 0$ and $\psi(\tau) \geq 0$, which are also always satisfied. Equation (6.21) with $n \geq 2$ provides nontrivial conditions. For example, the condition with $n = 2$ states that $\psi(\tau)$ monotonically decreases with τ.

We give two examples of $\psi(\tau)$ for which the Laplace Gillespie algorithm is applicable. First, consider the gamma distribution for $p(\lambda)$, i.e.,

$$p(\lambda) = \frac{\lambda^{\alpha-1} e^{-\lambda/\kappa}}{\Gamma(\alpha)\kappa^\alpha}, \tag{6.22}$$

where $\Gamma(\alpha)$ is the gamma function. By combining Eqs. (6.20) and (6.22), we obtain

$$\Psi(\tau) = \frac{1}{(1 + \kappa\tau)^\alpha}. \tag{6.23}$$

Differentiation of Eq. (6.23) leads to

$$\psi(\tau) = \frac{\kappa}{(1 + \kappa\tau)^{\alpha+1}}. \tag{6.24}$$

This is a power-law distribution.

Second, when the inter-event time obeys the gamma distribution, i.e.,

$$\psi(\tau) = \frac{\tau^{\alpha-1} e^{-\tau/\kappa}}{\Gamma(\alpha)\kappa^\alpha}, \tag{6.25}$$

$\Psi(\tau)$ is expressed as the Laplace transform of $p(\lambda)$ if and only if $0 < \alpha \leq 1$, where

$$p(\lambda) = \begin{cases} 0 & (0 < \lambda < \kappa^{-1}), \\ \frac{1}{\Gamma(\alpha)\Gamma(1-\alpha)\lambda(\kappa\lambda-1)^\alpha} & (\lambda \geq \kappa^{-1}). \end{cases} \tag{6.26}$$

6.3 Random walks

6.3.1 *Node-centric random walks*

We already saw examples of random walks on temporal networks, i.e., TempoRank (Section 4.5.1) and memory networks (Section 5.7). In these examples, the time was discrete and the snapshot representation was assumed.

In this section, we consider non-Markovian continuous-time random walks on static networks, which we regard as random walks on temporal networks. An application could be information diffusion on social networks in which the time for an agent to transfer the information after receiving it is described by a general renewal process. The event-based representation is assumed.

Consider a random walker in static networks in continuous time. By definition, the inter-event time, whose probability density function is $\psi(\tau)$, specifies the time between two jumps. Therefore, the inter-event time and the waiting time are equivalent in this model. To avoid an abuse of notation, we denote it by $\psi(\tau)$ rather than by $\psi^{\mathrm{w}}(\tau)$ throughout this section. This random walk model is node-centric in the sense that a renewal process is attached to each node. It is also called the active random walk [Hoffmann *et al.* (2012); Speidel *et al.* (2015)].

Assume that $\psi(\tau)$ is identical on each node for the sake of simplicity. We recall that the probability that the walker has moved n times at time t is denoted by $p(n, t)$. Equation (2.45) gives $p(n, t)$ in terms of ψ in the Laplace domain. By extending Eq. (2.65) to the case of general networks, we obtain

$$\boldsymbol{p}(t) = \sum_{n=0}^{\infty} \boldsymbol{p}(n) p(n, t), \qquad (6.27)$$

where $\boldsymbol{p}(t) = (p_1(t) \cdots p_N(t))$, $p_i(t)$ $(1 \leq i \leq N)$ is the probability that the walker is located at the ith node at time t, $\boldsymbol{p}(n) = (p_1(n) \cdots p_N(n))$ and $p_i(n)$ is the probability that the walker is located at the ith node after n moves. By Laplace transforming Eq. (6.27) and using Eqs. (2.45), we obtain

$$\hat{\boldsymbol{p}}(s) = \frac{1 - \hat{\psi}(s)}{s} \sum_{n=0}^{\infty} \boldsymbol{p}(n) \hat{\psi}(s)^n. \qquad (6.28)$$

By applying $\boldsymbol{p}(n) = \boldsymbol{p}(0) T^n$ (Eq. (2.108)) to Eq. (6.28), where T is the transition matrix of the discrete-time random walk on the network, we obtain

$$\hat{\boldsymbol{p}}(s) = \frac{1 - \psi(s)}{s} \boldsymbol{p}(0) \left[I - T \hat{\psi}(s) \right]^{-1}. \qquad (6.29)$$

Equation (6.29) generalises the results obtained by Montroll and Weiss [Montroll and Weiss (1965)] to arbitrary network structure. The inverse Laplace transform of Eq. (6.29) provides the probability $p_i(t)$ for the walker to be on the ith node at time t.

The dynamics are Markov only when jump events obey a Poisson process, i.e., when $\psi(\tau) = \lambda e^{-\lambda\tau}$. In this case, substitution of $\hat{\psi}(s) = \lambda/(s+\lambda)$ (Eq. (2.53)) in Eq. (6.29) leads after some calculations to

$$s\hat{p}(s) - p(0) = \lambda(-I + T)\hat{p}(s). \tag{6.30}$$

Because the inverse Laplace transform of $s\hat{p}(s) - p(0)$ is equal to $dp(t)/dt$, Eq. (6.30) results in a standard master equation given by

$$\frac{dp(t)}{dt} = \lambda(-I + T)p(t). \tag{6.31}$$

How does the shape of $\psi(\tau)$ affect the speed of diffusion? We can work out this problem in the "Fourier" domain, or working with the amplitude of the eigenmodes. Then, we will estimate the impact of the deviation from the Poisson process on the relaxation of the different eigenmodes [Delvenne *et al.* (2015)]. By combining Eqs. (3.43), (3.44) and (6.28), we obtain

$$\hat{p}(s) = \frac{1 - \hat{\psi}(s)}{s} \sum_{\ell=1}^{N} \frac{a_\ell(0)}{1 - \lambda_\ell \hat{\psi}(s)} u_\ell^{L}, \tag{6.32}$$

where λ_ℓ is an eigenvalue of the transition matrix, T, and u_ℓ^{L} is the corresponding left eigenvector. By taking the inner product of both sides of Eq. (6.32) with u_ℓ^{R}, where ℓ is a particular value, we obtain

$$\hat{a}_\ell(s) = \frac{1 - \hat{\psi}(s)}{s\left[1 - \lambda_\ell\hat{\psi}(s)\right]} a_\ell(0). \tag{6.33}$$

The relaxation of an eigenmode occurs exponentially in time for Poisson processes. However, different relaxation time courses are observed for non-exponential $\psi(\tau)$. To estimate the relaxation time of the eigenmodes, we assume that the inter-event time has a finite mean and variance and substitute the small s expansion (Eq. (2.76))

$$\hat{\psi}(s) = 1 - \langle\tau\rangle s + \frac{1}{2}\langle\tau^2\rangle s^2 + o(s^2) \tag{6.34}$$

in Eq. (6.33). For any decaying mode ℓ, i.e., those with $\lambda_\ell \neq 1$, calculations lead to

$$a_\ell(s) = \frac{\langle\tau\rangle}{1 - \lambda}\left[1 - s\left(\frac{\lambda_\ell\langle\tau\rangle}{1 - \lambda_\ell} + \frac{\langle\tau^2\rangle}{2\langle\tau\rangle}\right)\right]. \tag{6.35}$$

Equation (6.35) determines the characteristic time, denoted by t_{cha}, as follows:

$$t_{\text{cha}} = \frac{\lambda_\ell\langle\tau\rangle}{1 - \lambda_\ell} + \frac{\langle\tau^2\rangle}{2\langle\tau\rangle}$$

$$= \langle\tau\rangle\left(\frac{1}{\epsilon_\ell} + \beta\right), \tag{6.36}$$

where $\epsilon_\ell = 1 - \lambda_\ell$ is the eigenvalue of the normalised Laplacian, and

$$\beta = \frac{\sigma_\tau^2 - \langle\tau\rangle^2}{2\langle\tau\rangle^2}, \tag{6.37}$$

where $\sigma_\tau^2 = \langle\tau^2\rangle - \langle\tau\rangle^2$ is the variance of τ. β is a measure of burstiness and equal to $(\mathrm{CV}^2 - 1)/2$, where CV is the coefficient of variation (Section 4.6.2). Poisson processes yield $\beta = 0$. β ranges from $-1/2$, which is realised when $\psi(\tau)$ is a delta distribution, to arbitrarily large positive values for highly bursty activity.

As an illustration, let us consider the slowest non-stationary mode, usually called the mixing mode, associated with the spectral gap ϵ_ℓ, the smallest non-zero eigenvalue of the Laplacian. A small value of ϵ_ℓ implies the existence of eminent bottlenecks and communities in the underlying network. The characteristic decay time t_{cha} represents the worst-case relaxation time to the stationarity. Equation (6.36) presents competition between two factors. When the spectral gap is small, the first term of the right-hand side of Eq. (6.36) is dominant such that t_{cha} is mainly determined by the network structure. When the spectral gap is larger and hence the underlying network is less modular, or when burstiness is eminent, the second term is dominant such that t_{cha} is mainly governed by the properties of $\psi(\tau)$. This result shows that burstiness generally tends to slow down diffusion on networks. The present mechanism is different from the waiting-time paradox because the variance of the waiting-time distribution (denoted by $\psi(\tau)$ rather than $\psi^{\mathrm{w}}(\tau)$ in this section) appears in Eq. (6.36). It should be noted that, in the waiting-time paradox, the variance of the inter-event time, not of the waiting time, causes a paradox.

We can also calculate other properties of this random walk [Hoffmann *et al.* (2012); Speidel *et al.* (2015)]. Consider undirected networks for simplicity. The probability density that the walker transits from the ith to the jth nodes at time t since the walker has arrived at the ith node is given by

$$f(t; j \leftarrow i) = \psi(t) \left[\int_t^\infty \psi(t')\mathrm{d}t' \right]^{k_i - 1}, \tag{6.38}$$

where k_i is the degree of the ith node in the underlying static network. After some calculations, we obtain the stationary density as follows:

$$p_i^* = \frac{\langle\min_{\ell=1,\ldots,k_i} \tau_\ell\rangle k_i}{\sum_{j=1}^N \langle\min_{\ell=1,\ldots,k_j} \tau_\ell\rangle k_j}, \tag{6.39}$$

where τ_ℓ's are independent copies of the inter-event time. It should be noted that

$$\left\langle \min_{\ell=1,\ldots,k_i} \tau_\ell \right\rangle = \int_0^\infty \left[\int_t^\infty \psi(t')\mathrm{d}t' \right]^{k_i} \mathrm{d}t \tag{6.40}$$

depends only on k_i. Equations (6.39) and (6.40) indicate that p_i^* only depends on k_i. We can also calculate the mean recurrence time, i.e., the mean time needed for a walker starting at the ith node to return to the same node for the first time, denoted by $\langle T_{i|i} \rangle$. The mean recurrence time is given by

$$\langle T_{i|i} \rangle = \frac{\sum_{j=1}^{N} \langle \min_{\ell=1,\dots,k_j} \tau_\ell \rangle k_j}{k_i} \propto \frac{1}{k_i}. \qquad (6.41)$$

6.3.2 *Link-centric random walks*

Let us consider a different type of continuous-time random walk on temporal networks, i.e., one on stochastic temporal network model (Section 5.2). A random walker visiting node v moves to a different node once an event on a link involving node v occurs. This random walk model is link-centric in the sense that a renewal process is attached to each link. It is also called the passive random walk [Hoffmann *et al.* (2012); Speidel *et al.* (2015)].

This random walk is inherently non-Markovian except when the events are generated by Poisson processes. To see this, let us consider the probability of the time t at which the walker transits from v_i to v_j given that it transited from v_ℓ to v_i at time zero. The probability density of this event is denoted by $f(t; j \leftarrow i | i \leftarrow \ell)$. Assume that inter-event times of different links are identically and independently distributed according to $\psi(\tau)$. Starting from the time when the walker has moved from v_ℓ to v_i, the time of the next activation of link (v_i, v_ℓ), t, is given by $\psi(t)$. The time of the next activation of a different link (v_i, v_j) $(j \neq \ell)$ is determined by the waiting-time distribution, $\psi^{\mathrm{w}}(t)$, not by $\psi(t)$. This is because the point processes have been running on link (v_i, v_j) $(j \neq \ell)$ when the walker has arrived at v_i and we do not have information about the time of the last event on this link. For the walker to move to v_j at time t, no event must happen on all links but (v_i, v_j) in $[0, t]$. Therefore, we obtain

$$f(t; j \leftarrow i | i \leftarrow \ell) \approx \begin{cases} \psi(t) \left[\int_t^\infty \psi^{\mathrm{w}}(t')\mathrm{d}t' \right]^{k_i-1} & (j = \ell), \\ \psi^{\mathrm{w}}(t) \left[\int_t^\infty \psi^{\mathrm{w}}(t')\mathrm{d}t' \right]^{k_i-2} \int_t^\infty \psi(t')\mathrm{d}t' & (j \neq \ell). \end{cases}$$

$$(6.42)$$

Unless ψ is the exponential distribution, $f(t; \ell \leftarrow i | i \leftarrow \ell)$ is not equal to $f(t; j \leftarrow i | i \leftarrow \ell)$ $(j \neq \ell)$ in general, such that the trajectory of the random walk, i.e., walk measured in terms of the number of hops, is non-Markovian. If ψ is a long-tailed distribution, the waiting-time is larger than the inter-event time on average. Therefore, Eq. (6.42) implies that the walker tends to transit back to the node where it has come from (i.e., v_ℓ). Such non-Markovian trajectories emerge even if the activation times on different links are statistically independent. Even on randomised null models of temporal networks in which temporal correlations are removed, the tendency towards backtracking is expected to slow down diffusion dynamics. In contrast, trajectories are Markovian in the node-centric continuous-time random walk on networks, whereas that model is non-Markovian in the time domain.

Denote by $q_{j\leftarrow i}(t)$ the rate at which the walker reaches v_j from v_i at time t in the link-centric random walk. This quantity satisfies the following recursive relationship:

$$q_{j\leftarrow i}(t) \approx \sum_{\ell \in \mathcal{N}_i} \left[\int_0^t f(t - t'; j \leftarrow i | i \leftarrow \ell) q_{i\leftarrow\ell}(t')dt' \right] + p_{j\leftarrow i}(0)\delta(t), \quad (6.43)$$

where $\mathcal{N}_i$ is the set of the neighbours of v_i, the initial condition satisfies

$$\sum_{j \in \mathcal{N}_i} p_{i\leftarrow j}(0) = p_i(0), \quad (6.44)$$

and $\delta(t)$ is the delta function. Recall that $p_i(t)$ is the probability that the walker visits v_i at time t. Equation (6.44) implies that the initial condition requires not only the position of the walker but also where it has come from. This is because the transition probability in the next move depends on the node where the walker has come from. Finally, the master equation is given by

$$\frac{d}{dt}p_i(t) = \sum_{j \in \mathcal{N}_i} [q_{i\leftarrow j}(t) - q_{j\leftarrow i}(t)]. \quad (6.45)$$

To obtain the stationary density, it is more practical to work in terms of $q_{i\leftarrow j}(t)$ than $p_i(t)$. The Laplace transform of Eq. (6.43) is given by

$$\hat{q}_{j\leftarrow i}(s) \approx \sum_{\ell \in \mathcal{N}_i} \left[\hat{f}(s; j \leftarrow i | i \leftarrow \ell) \hat{q}_{i\leftarrow\ell}(s) \right] + p_{j\leftarrow i}(0). \quad (6.46)$$

Equation (6.46) is a set of linear equations involving $2M$ unknowns. Note that $\hat{q}_{j\leftarrow i}(s) \neq \hat{q}_{i\leftarrow j}(s)$ in general even for undirected temporal networks. We solve $\hat{q}_{j\leftarrow i}(s)$, which can be transformed back to $\lim_{t\to\infty} q_{j\leftarrow i}(t)$ and then to $p_i^* \equiv \lim_{t\to\infty} p_i(t)$. The final result is that $\lim_{t\to\infty} q_{j\leftarrow i}(t)$ is independent of i and j and that

$$p_i^* = \frac{1}{N} \quad (1 \leq i \leq N). \quad (6.47)$$

Therefore, the stationary density is the uniform density irrespectively of the underlying network and $\psi(\tau)$. We can also calculate the mean recurrence time, which is given by

$$\langle T_{i|i} \rangle \approx \frac{N\langle \tau \rangle}{k_i}. \tag{6.48}$$

Equation (6.48) indicates that the mean recurrence time depends on $\psi(\tau)$ only through the mean inter-event time, $\langle \tau \rangle$.

6.4 Epidemic processes

Epidemic processes are probably the most studied dynamical processes on networks, both for static and temporal networks. Many of these investigations are motivated by their applications to infectious diseases of humans and animals, viral and other information spreading on social networks, and computer viruses. In contrast to a wealth of theory available for static networks [Newman (2010); Pastor-Satorras *et al.* (2015)], studies of epidemic processes on temporal networks are more often numerically conducted, presumably because of the difficulty of the problem. In this section, we first explain three representative models of epidemic dynamics on networks or well-mixed populations (Section 6.4.1). Then, we introduce four models of epidemic spreading on temporal networks, each developing different modelling techniques.

6.4.1 *Models of epidemic processes*

The susceptible-infected-susceptible (SIS) model, the susceptible-infected-recovered (SIR) model and the susceptible-infected (SI) models are probably the most frequently studied epidemic processes in complex networks. These models are named after the types of state that each node assumes, i.e., the susceptible (in short, healthy), infected and recovered states, and admitted transitions between the states. They are called compartmental models, where compartment is a synonym of state. We focus on stochastic versions of these models, which are usually studied when considered on networks. The transition rates of the three models, which fully define the models, are summarised in Fig. 6.3(a).

The SIS model assumes two processes. When a susceptible node interacts with an infected node, the susceptible node contracts infection to transit to the infected state at rate β. In other words, the probability that

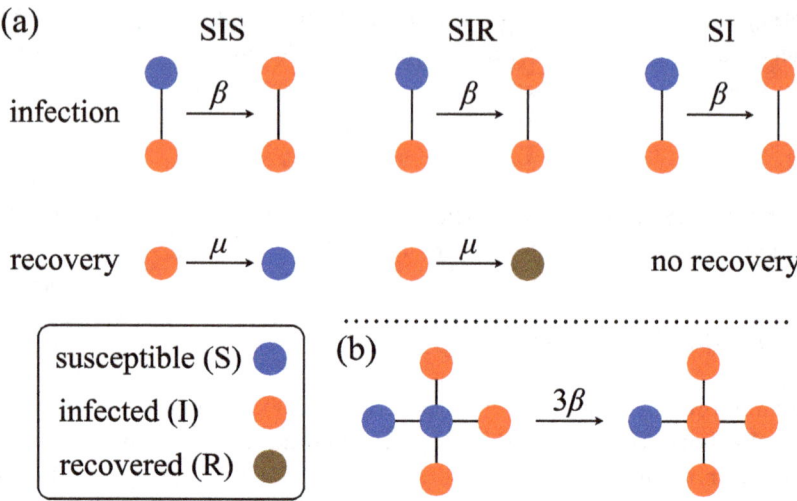

Fig. 6.3 (a) Infection and recovery rates in three epidemic process models. (b) Infection rate when a susceptible node is surrounded by one susceptible node and three infected nodes.

a susceptible node gets infected in small time Δt is equal to $\beta \Delta t$. If a susceptible node is adjacent to k_I infected neighbours, the transition rate is equal to $k_I \beta$ (see Fig. 6.3(b) for an example of $k_I = 3$). An infected node recovers at rate μ irrespectively of the states of the neighbours. Once an infected node recovers, it transits to the susceptible state. Therefore, a node may contract infection multiple times during a single run of the SIS model.

Consider the mean-field population, i.e., the complete graph, where each node pair is connected with each other with a normalised weight of $1/N$. Denote the fraction of susceptible and infected nodes at time t by $S(t)$ and $I(t)$, respectively. The dynamics in the thermodynamic limit (i.e., $N \to \infty$) are given by

$$\frac{dS(t)}{dt} = -\beta I(t)S(t) + \mu I(t) \qquad (6.49)$$

and

$$\frac{dI(t)}{dt} = \beta S(t)I(t) - \mu I(t). \qquad (6.50)$$

The first terms in Eqs. (6.49) and (6.50) represent the fact that each of the $NS(t)$ susceptible nodes is infected at a total rate of $\beta I(t)$. Addition of Eqs. (6.49) and (6.50) yields

$$\frac{d}{dt}\{S(t) + I(t)\} = 0. \qquad (6.51)$$

Therefore, $S(t) + I(t) = S(0) + I(0) = 1$ such that the total number of nodes is conserved.

In the equilibrium, setting the right-hand side of Eq. (6.50) to zero yields

$$I^*(\beta S^* - \mu) = 0, \tag{6.52}$$

where S^* and I^* are the fraction of susceptible and infected nodes in the equilibrium, respectively. If infection persists in the population in the equilibrium, by applying $I^* \neq 0$ and $S^* = 1 - I^*$ to Eq. (6.52), we obtain

$$I^* = 1 - \frac{\mu}{\beta}. \tag{6.53}$$

Equation (6.53) is intuitive in that the magnitude of infection is large if the infection rate β is large or the recovery rate μ is small. In addition, $I^* > 0$ holds true if and only if

$$\frac{\beta}{\mu} > 1. \tag{6.54}$$

We say that the epidemic threshold in terms of β/μ is equal to unity in the well-mixed population. In fact, the epidemic threshold depends on the structure of the underlying network. For example, the configuration model (Section 3.8.2) yields the epidemic threshold equal to $\langle k \rangle / \langle k^2 \rangle$ [Newman (2010); Pastor-Satorras *et al.* (2015)]. For scale-free networks, $\langle k^2 \rangle$ is very large such that the epidemic threshold can be extremely small.

In the SIR model, infection events occur in the same manner as in the SIS model. The only difference to the SIS model is that when an infected node recovers at rate μ, it transits to the recovered state, not back to the susceptible state. A recovered node does not infect others or is not reinfected. The recovered state can also be interpreted as the removed or dead state because a dead node would not infect or be infected by others. In contrast to the SIS model, infectious nodes are eventually extinct in the SIR model even if the infection rate is high. For an arbitrary initial condition, the final state consists of susceptible and recovered nodes, but not infected nodes. We typically start the SIR model from a single infected node or a small fraction of infected nodes in the background of the susceptible population. The primary interest is in the final size, i.e., the number of recovered individuals when the dynamics have terminated.

The SIR model is suitable for describing the response of a population to a triggering event, such as the viral spreading of a tweet in Twitter. Because such one-shot epidemic dynamics are relevant to many real phenomena, the SIR model and its variants are probably more frequently used than the SIS model unless a slow time scale set by births and deaths of individuals comes into play; birth and death events make the SIR model similar to extensions of the SIS model.

The SIR dynamics for the well-mixed population are described by

$$\frac{dS(t)}{dt} = -\lambda I(t)S(t), \tag{6.55}$$

$$\frac{dI(t)}{dt} = \lambda I(t)S(t) - \mu I(t), \tag{6.56}$$

$$\frac{dR(t)}{dt} = \mu I(t). \tag{6.57}$$

Summation of Eqs. (6.55), (6.56) and (6.57) confirms that the total number of nodes is conserved, i.e., $S(t) + I(t) + R(t) = 1$ for all t.

When $dI(t)/dt > 0$ at $t = 0$, the number of infectious nodes first increases to a macroscopic (i.e., $O(N)$) number. In this case, we regard that an outbreak has occurred. Otherwise, initially infected nodes do not trigger secondary infections on a visible scale. The condition $dI(t)/dt|_{t=0} > 0$ gives the epidemic threshold, and the result coincides with that for the SIS model given by Eq. (6.54). For heterogeneous networks, the epidemic threshold in terms of β/μ is given by $\langle k \rangle / (\langle k^2 \rangle - \langle k \rangle)$ [Newman (2010); Pastor-Satorras *et al.* (2015)]. Although this expression is slightly different from the epidemic threshold for the SIS model, scale-free networks, which yield $\langle k^2 \rangle \gg \langle k \rangle$, lead to a very small epidemic threshold in both SIS and SIR models.

The third model, the SI model, is a simplified version of the SIR model. Infection events occur in the same manner as in the SIS and SIR models. In the SI model, once a node is infected, it will stay infected forever. Therefore, if infection is introduced to a connected static network, every node will be eventually infected. No notion of epidemic threshold exists for the SI model. Instead, one is interested in how fast infection spreads. The dynamics in the well-mixed population are governed by

$$\frac{dI(t)}{dt} = \lambda S(t)I(t), \tag{6.58}$$

where we omit the equation for the dynamics of $S(t)$. In an early stage of dynamics where only a tiny fraction of nodes is infected, Eq. (6.58) with $S(t) \approx 1$ yields

$$I(t) \propto e^{\lambda t}. \tag{6.59}$$

If the observation period is finite, as is typical in temporal network data, some nodes may be susceptible at the final time such that the fraction of infected nodes at the final time is a relevant question to the SI model. Although the SI model is unrealistic, it behaves similarly to the SIR model in the initial stage of the dynamics, where the number of recovered nodes in the SIR model can be safely neglected. The SI model is relatively well employed in temporal rather than static network studies, presumably because temporal networks are already complex objects. The use of the SI as opposed to the SIR model suppresses the complexity of the entire model (i.e., an epidemic process model on temporal networks).

6.4.2   *SIS dynamics on metapopulation models*

In this and the following sections, we will explain analysis of four epidemic process models in temporal networks. In this section, we derive the epidemic threshold for the SIS model on metapopulation models [Colizza *et al.* (2007); Masuda (2010)]. Although metapopulation models are not usually discussed as temporal networks, this example tells us how mobility of individuals interacts with dynamics of infection and recovery. For analysis of the SIR model on metapopulation models, see [Colizza and Vespignani (2008); Balcan and Vespignani (2011)]. Here we present the SIS rather than the SIR model because the theory is simpler.

The master equation for the number of individuals in each subpopulation is given by Eq. (5.69) when a epidemic process does not take place. We denote by $N_{\mathrm{S},i}$ and $N_{\mathrm{I},i}$ the numbers of susceptible and infected nodes in the ith subpopulation, respectively. We assume that the diffusion rates for susceptible and infected individuals, D_{S} and D_{I}, respectively, are possibly different and that each pair of individuals in the same subpopulation interacts at a constant rate. The master equations are given by

$$\frac{\mathrm{d}N_{\mathrm{S},i}}{\mathrm{d}t} = -\beta N_{\mathrm{S},i}N_{\mathrm{I},i} + \mu N_{\mathrm{I},i} - D_{\mathrm{S}}N_{\mathrm{S},i} + D_{\mathrm{S}}\sum_{j=1}^{\tilde{N}}\frac{\tilde{A}_{ji}}{\tilde{k}_j}N_{\mathrm{S},j}, \qquad (6.60)$$

$$\frac{\mathrm{d}N_{\mathrm{I},i}}{\mathrm{d}t} = \beta N_{\mathrm{S},i}N_{\mathrm{I},i} - \mu N_{\mathrm{I},i} - D_{\mathrm{I}}N_{\mathrm{I},i} + D_{\mathrm{I}}\sum_{j=1}^{\tilde{N}}\frac{\tilde{A}_{ji}}{\tilde{k}_j}N_{\mathrm{I},j}, \qquad (6.61)$$

where $\tilde{N}$ is the number of subpopulations such that $1 \leq i \leq \tilde{N}$, $\tilde{A}$ is the adjacency matrix of the network of the subpopulations, and $\tilde{k}_i = \sum_{j=1}^{\tilde{N}}\tilde{A}_{ij}$ is the degree of the ith subpopulation.

To determine the epidemic threshold, we consider the situation in which $N_{\mathrm{I},i}$ $(1 \leq i \leq N)$ is infinitesimally small. In this situation, Eq. (6.60) is essentially the same as the master equation without epidemic dynamics

(Eq. (5.69)) because $\beta N_{S,i} N_{I,i}$ and $\mu N_{I,i}$ on the right-hand side of Eq. (6.60) are very small as compared to the other terms. As long as susceptible individuals move (i.e., $D_S > 0$), the equilibrium for the susceptible individuals is given by

$$N_{S,i}^* = \frac{\tilde{k}_i N}{\langle \tilde{k} \rangle \tilde{N}}, \qquad (6.62)$$

where N is the number of individuals in the entire population. Equation (6.62) is the same as Eq. (5.71). Substitution of Eq. (6.62) in Eq. (6.61) yields

$$\frac{dN_{I,i}}{dt} \approx \frac{\beta N}{\langle \tilde{k} \rangle \tilde{N}} \tilde{k}_i N_{I,i} - \mu N_{I,i} - D_I N_{I,i} + D_I \sum_{j=1}^{\tilde{N}} \frac{\tilde{A}_{ji}}{\tilde{k}_j} N_{I,j}. \qquad (6.63)$$

Equation (6.63) admits the disease-free solution $N_{I,i}^* = 0$ ($1 \leq i \leq N$). The destabilisation of the disease-free solution implies an endemic state, where infection can persist. We rewrite Eq. (6.63) in the vector form:

$$\frac{d\boldsymbol{N}_I}{dt} \approx B\boldsymbol{N}_I, \qquad (6.64)$$

where $\boldsymbol{N}_I = (N_{I,1}, \ldots, N_{I,\tilde{N}})^\top$ and B is the $\tilde{N} \times \tilde{N}$ matrix whose elements are given by

$$B_{ij} = \delta_{ij} \left(\frac{\beta N}{\langle \tilde{k} \rangle \tilde{N}} \tilde{k}_i - \mu - D_I \right) + (1 - \delta_{ij}) D_I \frac{\tilde{A}_{ji}}{\tilde{k}_j}. \qquad (6.65)$$

The epidemic threshold is given when the largest eigenvalue of B crosses zero from below as β increases (Section 2.8). In contrast to the case of mean-field populations and static networks (Section 6.4.1), the epidemic threshold is not expressed only in terms of β/μ.

Let us look at some special cases. When $D_I = \infty$, Eq. (6.63) implies

$$N_{I,i} = \frac{\tilde{k}_i N_I}{\langle \tilde{k} \rangle \tilde{N}}, \qquad (6.66)$$

where $N_I \equiv \sum_{i=1}^{\tilde{N}} N_{I,i} \approx 0$ is the number of infected individuals in the entire population near the epidemic threshold. By substituting Eq. (6.66) in Eq. (6.63) and taking the summation over i, we obtain

$$\frac{dN_I}{dt} = \left(\frac{\beta N \langle \tilde{k}^2 \rangle}{\langle \tilde{k} \rangle^2 \tilde{N}} - \mu \right) N_I, \qquad (6.67)$$

where $\langle \tilde{k}^2 \rangle \equiv \sum_{i=1}^{\tilde{N}} \tilde{k}_i^2 / \tilde{N}$. Therefore, the condition for endemicity is given by

$$\frac{\beta}{\mu} > \frac{\langle \tilde{k} \rangle^2 \tilde{N}}{\langle \tilde{k}^2 \rangle N}. \tag{6.68}$$

A scale-free metapopulation model network yields a small epidemic threshold because $\langle \tilde{k}^2 \rangle \gg \langle \tilde{k} \rangle$.

When $D_I = 0$, Eq. (6.63) represents $\tilde{N}$ decoupled dynamics. The epidemic threshold is determined by the subpopulation having the largest degree, denoted by $\tilde{k}_{\max}$. The condition for endemicity is given by

$$\frac{\beta}{\mu} > \frac{\langle \tilde{k} \rangle \tilde{N}}{\tilde{k}_{\max} N}. \tag{6.69}$$

In fact, even if this condition is met, only the subpopulations having the largest degrees accommodate infected individuals.

6.4.3 *SIR dynamics on the neighbour exchange network model*

In [Volz and Meyers (2007)], the authors semi-analytically examined the dynamics of the SIR model on a temporal network model. Their model of temporal network, called the neighbour exchange model, is driven by link swapping. Assume an undirected network. Two randomly chosen nodes exchange one of their neighbours at a constant rate ρ. A neighbour exchange event involves four nodes and does not affect the degree of any of them (Fig. 6.4). Therefore, the degree of each node is preserved over time, whereas their neighbours vary over time. The model allows an arbitrary degree distribution. On top of the neighbour exchange model, the SIR dynamics are assumed to occur, typically starting from a single infected individual. The occurrence rate of the neighbour exchange events is independent of the state of nodes (i.e., susceptible, infected or recovered).

The parameter ρ controls the temporality of the network. In the limit $\rho \to \infty$, the network blinks very fast as compared to the time scale of epidemic. In this situation, the network is essentially annealed, with every pair of nodes being adjacent to each other with a properly scaled link weight. As we decrease ρ from $\rho \to \infty$, the network becomes gradually temporal, whereby the time scale of network changes and that of epidemic dynamics gradually become comparable.

If we discard the stochasticity in the dynamics, the model can be analysed by the deterministic formalism, i.e., a set of differential equations.

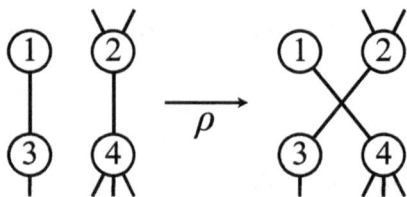

Fig. 6.4   Dynamics of neighbour exchange. Two links are exchanged at rate ρ.

In fact, the final size increases as ρ increases [Volz and Meyers (2007)]. They reached this conclusion by numerically simulating a set of differential equations, whose accuracy was confirmed by their direct numerical simulations of the stochastic dynamics of the model. Therefore, the epidemic is hindered by temporal networks (i.e., finite ρ) as compared to aggregate networks (i.e., $\rho \to \infty$). Their analysis exploits the generating functions to account for heterogeneous degree distributions and is too involved to be presented here. Therefore, we restrict ourselves to the case of regular networks, i.e., networks in which all nodes have the same degree k, and conduct a simplified analysis.

Let us start with the case without neighbour exchange, i.e., $\rho = 0$ [Volz (2008)]. We are interested in the dynamics of the fraction of susceptible individuals, $S(t)$, and that of infected individuals, $I(t)$, where t denotes the time. The fraction of recovered individuals is given by $1 - S(t) - I(t)$. If we take into account the correlation between the states of the adjacent nodes, we obtain

$$\frac{\mathrm{d}S(t)}{\mathrm{d}t} = -\,\beta S(t)kp_{\mathrm{I}}(t), \tag{6.70}$$

$$\frac{\mathrm{d}I(t)}{\mathrm{d}t} = \beta S(t)kp_{\mathrm{I}}(t) - \mu I(t), \tag{6.71}$$

where $p_{\mathrm{I}}(t)$ is the probability that a link with a susceptible node on one end has an infected node on the other end. This quantity is not generally equal to $I(t)$ due to the correlation between nodes. The rate of infection events is accounted for by $\beta S(t)kp_{\mathrm{I}}(t)$, and that of recovery events by $\mu I(t)$.

The dynamics of $p_{\mathrm{I}}(t)$ are given by

$$\frac{\mathrm{d}p_{\mathrm{I}}(t)}{\mathrm{d}t} = \frac{\mathrm{d}}{\mathrm{d}t}\frac{M_{\mathrm{SI}}(t)}{S(t)} = \frac{\frac{\mathrm{d}M_{\mathrm{SI}}(t)}{\mathrm{d}t}S(t) - M_{\mathrm{SI}}(t)\frac{\mathrm{d}S(t)}{\mathrm{d}t}}{S(t)^2}, \tag{6.72}$$

where $M_{\mathrm{SI}}(t)$ is the fraction of links whose endpoints are a susceptible node and an infected node at time t.

Equation (6.72) indicates that we need to know $\mathrm{d}M_{\mathrm{SI}}(t)/\mathrm{d}t$ to derive $\mathrm{d}p_{\mathrm{I}}(t)/\mathrm{d}t$; $\mathrm{d}S(t)/\mathrm{d}t$ is given by Eq. (6.70). To derive $\mathrm{d}M_{\mathrm{SI}}(t)/\mathrm{d}t$, let us consider a pair of adjacent nodes denoted by v_i and v_j. $M_{\mathrm{SI}}(t)$ increases if both v_i and v_j are susceptible and v_j is infected by one of its $k-1$ neighbours excluding v_i. This event occurs at rate $S(t)p_{\mathrm{S}}(t) \times \beta(k-1)p_{\mathrm{I}}(t)$. The first factor $S(t)p_{\mathrm{S}}(t)$ represents the probability that v_i is susceptible with probability $S(t)$ and v_i has a susceptible neighbour, which we regard to be v_j, with probability $p_{\mathrm{S}}(t)$. Here, $p_{\mathrm{S}}(t)$ is defined as the probability that a link possessing a susceptible node on one end has a susceptible neighbour on the other end. The second factor $\beta(k-1)p_{\mathrm{I}}(t)$ represents the rate at which v_j contracts infection. $M_{\mathrm{SI}}(t)$ decreases if v_i is susceptible, v_j is infected, and either of the following three events occurs. First, v_i can get infected by v_j. This event occurs at rate $\beta M_{\mathrm{SI}}(t)$. Second, v_j can recover. This event occurs at rate $\mu M_{\mathrm{SI}}(t)$. Third, v_i can get infected by an infected neighbour different from v_j. This event occurs at a rate of $S(t)p_{\mathrm{I}}(t) \times \beta(k-1)p_{\mathrm{I}}(t)$. By collecting all the contributions, we obtain

$$\frac{\mathrm{d}M_{\mathrm{SI}}(t)}{\mathrm{d}t} = \beta(k-1)S(t)p_{\mathrm{I}}(t)\left[p_{\mathrm{S}}(t) - p_{\mathrm{I}}(t)\right] - (\beta + \mu)M_{\mathrm{SI}}(t). \qquad (6.73)$$

By substituting Eqs. (6.70), (6.73) and $M_{\mathrm{SI}}(t) = S(t)p_{\mathrm{I}}(t)$ in Eq. (6.72), we obtain

$$\frac{\mathrm{d}p_{\mathrm{I}}(t)}{\mathrm{d}t} = \beta(k-1)p_{\mathrm{S}}(t)p_{\mathrm{I}}(t) - \beta p_{\mathrm{I}}(t)\left[1 - p_{\mathrm{I}}(t)\right] - \mu p_{\mathrm{I}}(t). \qquad (6.74)$$

The dynamics of $p_{\mathrm{S}}(t)$ are derived in a manner similar to the derivation of $p_{\mathrm{I}}(t)$. In terms of $M_{\mathrm{SS}}(t)$ defined as the fraction of links both of whose endpoints are susceptible nodes, we obtain

$$\frac{\mathrm{d}p_{\mathrm{S}}(t)}{\mathrm{d}t} = \frac{\mathrm{d}}{\mathrm{d}t}\frac{M_{\mathrm{SS}}(t)}{S(t)} = \frac{\frac{\mathrm{d}M_{\mathrm{SS}}(t)}{\mathrm{d}t}S(t) - M_{\mathrm{SS}}(t)\frac{\mathrm{d}S(t)}{\mathrm{d}t}}{S(t)^2}. \qquad (6.75)$$

Assume that both v_i and v_j are susceptible. $M_{\mathrm{SS}}(t)$ only decreases, and it does so when v_j is infected by a neighbour different from v_i. Therefore, we obtain

$$\frac{\mathrm{d}M_{\mathrm{SS}}(t)}{\mathrm{d}t} = -2\beta(k-1)S(t)p_{\mathrm{S}}(t)p_{\mathrm{I}}(t). \qquad (6.76)$$

The factor 2 on the right-hand side accounts for the fact the role of v_i and v_j can be exchanged in the arguments leading to Eq. (6.76). By substituting Eqs. (6.70), (6.76) and $M_{\mathrm{SS}}(t) = S(t)p_{\mathrm{S}}(t)$ in Eq. (6.75), we obtain

$$\frac{\mathrm{d}p_{\mathrm{S}}(t)}{\mathrm{d}t} = -2\beta(k-2)p_{\mathrm{S}}(t)p_{\mathrm{I}}(t). \qquad (6.77)$$

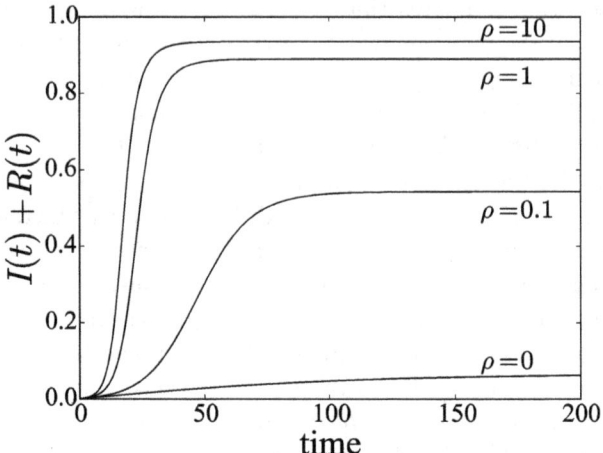

Fig. 6.5   SIR dynamics in the neighbour exchange model. We set $\beta = \mu = 0.2$, $k = 3$, $S(0) = p_S(0) = 0.999$ and $I(0) = p_I(0) = 0.001$ for each value of ρ.

The entire SIR dynamics on the static regular network are described by Eqs. (6.70), (6.71), (6.74) and (6.77).

Now, we extend the analysis to the case of the neighbour exchange model. Suppose that node v_i is susceptible and has an infected neighbour v_j, which contributes to $p_I(t)$. By a link swapping event, v_i severs the link to v_j, thus decreasing $p_I(t)$, and instead connects to a node selected as an endpoint of a randomly selected link. The probability that the new neighbour is infected is equal to $I(t)$. Therefore, we modify Eq. (6.74) to

$$\frac{dp_I(t)}{dt} = \beta(k-1)p_S(t)p_I(t) - \beta p_I(t)\left[1 - p_I(t)\right] - \mu p_I(t) + \rho\left[I(t) - p_I(t)\right].$$
(6.78)

Similarly, we extend Eq. (6.77) to

$$\frac{dp_S(t)}{dt} = -2\beta(k-2)p_S(t)p_I(t) + \rho\left[S(t) - p_S(t)\right].$$
(6.79)

The numerical results for the set of differential equations Eqs. (6.70), (6.71), (6.78) and (6.79) are shown in Fig. 6.5. The size of epidemic, $I(t) + R(t)$, increases as ρ increases, which is consistent with the results shown in [Volz and Meyers (2007)], and also with some numerical results on empirical temporal networks and their null models (reviewed in [Masuda and Holme (2013)]).

In the limit of the aggregate network, i.e., $\rho \to \infty$, Eqs. (6.78) and (6.79) imply $I(t) = p_I(t)$ and $S(t) = p_S(t)$, respectively. Substitution of these equalities in Eqs. (6.70) and (6.71) yields

$$\frac{dS(t)}{dt} = -\beta k S(t) I(t),$$ (6.80)

$$\frac{dI(t)}{dt} = \beta k S(t) I(t) - \mu I(t).$$ (6.81)

Equations (6.80) and (6.81) are the SIR model in the mean-field population such that each node pair is adjacent with link weight k/N. As ρ decreases from infinity, the time scale of the network dynamics becomes comparable with that of the SIR dynamics. Figure 6.5 indicates that the temporality of the network inhibits epidemic spreading. When $\rho = 0$, we again have a static network, i.e., a particular realisation of the regular random graph with degree k. However, this network is not comparable with temporal networks (i.e., a positive value of ρ) in the sense that the aggregation (i.e., time average) of the latter, which is the mean-field population, is not equal to the static network obtained at $\rho = 0$.

The neighbour exchange model is implicitly built on the event-based representation of temporal networks but not concerned with statistics of events such as burstiness. In the next section, we introduce another theoretical approach to epidemic dynamics on event-based representations of temporal networks, which explicitly considers the statistics of inter-event times.

6.4.4 *Viral spreading dynamics under bursty interaction*

Information often travels very slowly in social networks. If we model information spreading by the SI or SIR model on static networks, the number of infected nodes, corresponding to those who have caught new information, would grow and then decay relatively quickly in time. However, data on computer viruses [Vazquez *et al.* (2007)] and viral marketing campaign experiments [Iribarren and Moro (2009, 2011)] show that the number of newly infected nodes decays very slowly obeying long-tailed distributions. In this section, we deal with the Bellman-Harris branching processes model, which generalises the Galton-Watson processes (Section 2.10) and accounts for long-tailed behaviour of information spreading [Iribarren and Moro (2009, 2011)]. The model is a variant of the SIR model on an infinite tree. It discards explicit structure of networks and instead focuses on the effects of long-tailed distributions of inter-event times on spreading dynamics. For other closely related theoretical approaches, see [Jo *et al.* (2014)] for a different branching process model and [Karrer and Newman (2010)] for a so-called message passing approach.

Imagine a viral cascade starting from a single seed node. The seed node is assumed to become spreader with probability λ_0. When it does, the seed node simultaneously sends the message to r_0 others and then stops spreading the message. A spreader corresponds to the infected individual, and one that has stopped spreading the message corresponds to the recovered individual in the SIR model. A node that has received the message independently becomes secondary spreader with probability λ_1. A secondary spreader simultaneously spreads the message to r_1 other nodes just once.

The mean of r_0 and r_1 is denoted by

$$\overline{r}_i \equiv \sum_{r=0}^{\infty} r p_{i,r} \quad (i = 0, 1), \tag{6.82}$$

where $p_{i,r}$ is the probability that $r_i = r$. Empirical evidence supports $\overline{r}_0 < \overline{r}_1$ for two reasons. First, a seed node finds the campaign by chance, whereas a second spreader receives the message from a reliable source such as a friend. Therefore, a secondary spreader is considered to be more motivated to spread the message than the seed is. Second, because of the friendship paradox (Section 3.2), the average degree of a secondary spreader should be larger than that of the seed node. Therefore, we distinguish spreading from the seed node and that from a secondary spreader.

The primary and secondary basic reproduction numbers, denoted by R_0 and R_1, respectively, represent the mean number of infections caused by the corresponding spreader. They are given by

$$R_i \equiv \lambda_i \overline{r}_i = \lambda_i \sum_{r=0}^{\infty} r p_{i,r} \quad (i = 0, 1). \tag{6.83}$$

We define the generating functions for the number of recommendations sent by a node by

$$f_i(x) = 1 - \lambda_i + \lambda_i \sum_{r=0}^{\infty} x^r p_{i,r} \quad (i = 0, 1), \tag{6.84}$$

which yield $R_i = f_i'(1)$ $(i = 0, 1)$.

Denote by τ_0 and τ_1 the response times for the seed and a secondary spreader, respectively. They are allowed to obey different distributions whose cumulative distribution functions are denoted by $G_0(\tau)$ and $G_1(\tau)$, respectively. These response times typically obey long-tailed distributions.

The number of nodes that are spreading the message at time t is denoted by $I_0(t)$ and $I_1(t)$ depending on whether the entire spreading dynamics have started from a seed node or a secondary spreader node, respectively, at $t = 0$. To facilitate analysis using branching processes, we put the following crucial assumptions. First, when a spreader spreads the message to r_0 or r_1 others, those who have received the message, i.e., newly recruited secondary spreaders, start spreading the message as independent and identical copies of $I_1(t)$. Second, we ignore the structure of the social network. It should be noted that, even if r_1 is heterogeneously distributed according to distribution $p_{1,r}$, a node with large r_1 does not necessarily have a large degree. This is because r_1 (and also r_0) may be affected by non-network factors. Third, we assume that the population is infinite. In fact, the population is finite such that a spreader would find susceptible nodes with more difficulty as the dynamics progress. Although this fact would decrease r_1 over time, we neglect it.

Under these assumptions, we obtain

$$I_1(t) = \begin{cases} 1 & (t < \tau), \\ \sum_{i=1}^{r_1} I_1^{(i)}(t - \tau) & (t \geq \tau), \end{cases} \tag{6.85}$$

where $I_1^{(i)}(t)$ are independent and identically distributed copies of $I_1(t)$. Equation (6.85) indicates that the number of infected nodes, i.e., those actively spreading the message, remains unity, with the initial secondary spreader being the only one, until it propagates the message to r_1 others at $t = \tau$. Afterwards, the initial secondary spreader ceases to spread the message. An entire process is schematically shown in Fig. 6.6.

In terms of the generating function defined by

$$F_1(s, t) \equiv \sum_{j=0}^{\infty} p\left[I_1(t) = j\right] s^j, \tag{6.86}$$

where p denotes the probability as usual, we obtain

$$F_1(s, t) = \begin{cases} s & (t < \tau), \\ f_1\left[F_1(s, t - \tau)\right] & (t \geq \tau). \end{cases} \tag{6.87}$$

Because the inter-event time τ obeys the cumulative distribution function $G_1(\tau)$, we can rewrite Eq. (6.87) as

$$F_1(s, t) = s\left[1 - G_1(t)\right] + \int_0^t f_1\left[F_1(s, t - \tau)\right] dG_1(\tau). \tag{6.88}$$

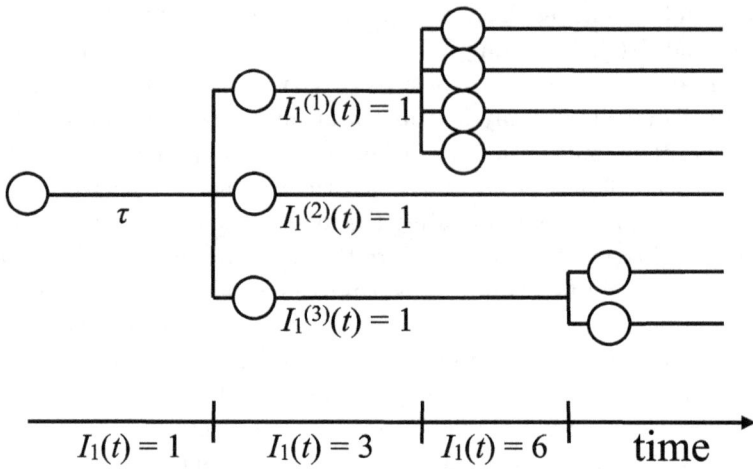

Fig. 6.6   Schematic of the Bellman-Harris branching process starting from a secondary spreader.

It should be noted that $dG_1(\tau)$ is essentially the same as the probability density function of inter-event times multiplied by $d\tau$. The generating function for $I_0(t)$, i.e., $F_0(s,t) \equiv \sum_{j=0}^{\infty} p\,[I_0(t) = j]\,s^j$ is similarly given by

$$F_0(s,t) = s\,[1 - G_0(t)] + \int_0^t f_0\,[F_1(s,t-\tau)]\,dG_0(\tau). \qquad (6.89)$$

A system of equations similar to Eqs. (6.88) and (6.89) frequently appears in generating function analysis of network phenomena such as percolation and epidemic processes. One first has to solve Eq. (6.88) for $F_1(s,t)$ and then insert the solution to Eq. (6.89) to solve for $F_0(s,t)$.

We can calculate the size of the cascade at time t starting from an initial seed and secondary spreader, denoted by $S_0(t)$ and $S_1(t)$, respectively. They are the number of nodes that have received the message up to time t. Roughly speaking, $S_0(t)$ and $S_1(t)$ are "integrals" of $I_0(t)$ and $I_1(t)$ over time. We obtain

$$S_0(t) = \begin{cases} 1 & (t < \tau), \\ 1 + \sum_{i=1}^{r_0} S_1^{(i)}(t-\tau) & (t \geq \tau), \end{cases} \qquad (6.90)$$

where

$$S_1(t) = \begin{cases} 1 & (t < \tau), \\ 1 + \sum_{i=1}^{r_1} S_1^{(i)}(t-\tau) & (t \geq \tau). \end{cases} \qquad (6.91)$$

In terms of the generating functions, we obtain

$$\Phi_0(s,t) \equiv \sum_{j=1}^{\infty} p\left[S_0(t) = j\right] s^j$$

$$= s\left[1 - G_0(t)\right] + s \int_0^t f_0\left[\Phi_1(s, t - \tau)\right] dG_0(\tau), \qquad (6.92)$$

$$\Phi_1(s,t) \equiv \sum_{j=1}^{\infty} p\left[S_1(t) = j\right] s^j$$

$$= s\left[1 - G_1(t)\right] + s \int_0^t f_1\left[\Phi_1(s, t - \tau)\right] dG_1(\tau). \qquad (6.93)$$

Equations (6.92) and (6.93) in the limit $t \to \infty$ read

$$\Phi_0(s, t \to \infty) = s f_0\left[\Phi_1(s, t \to \infty)\right], \qquad (6.94)$$

$$\Phi_1(s, t \to \infty) = s f_1\left[\Phi_1(s, t \to \infty)\right], \qquad (6.95)$$

respectively, because $G_0(t)$ and $G_1(t)$ are cumulative distributions of inter-event times, which tend to unity as $t \to \infty$. The average size of cascades is asymptotically given by $\langle S_0(t \to \infty)\rangle = \Phi_0'(1, t \to \infty)$. By differentiating Eq. (6.95) and setting $s = 1$, we obtain

$$\Phi_1'(1, t \to \infty) = 1 + R_1 \Phi_1'(1, t \to \infty), \qquad (6.96)$$

where we used $f_1\left[\Phi_1(1, t \to \infty)\right] = f_1(1) = 1$ and $f_1'\left[\Phi_1(1, t \to \infty)\right] = f_1'(1) = R_1$. By differentiating Eq. (6.94), setting $s = 1$ and substituting Eq. (6.96), we obtain

$$\langle S_0(t \to \infty)\rangle = 1 + \frac{R_0}{1 - R_1}, \qquad (6.97)$$

where we used $f_0'\left[\Phi_1(1, t \to \infty)\right] = f_0'(1) = R_0$. Equation (6.97) indicates that the average cascade size diverges as $R_1 \to 1$. This result implies that the spreading tree explodes with some probability. Otherwise, the spreading will terminate eventually. This result is intuitive because R_1 is the mean number of secondary spreaders that a secondary spreader recruits.

The generating function formalism of the Bellman-Harris branching processes also informs us of temporal dynamics of cascade processes. The mean dynamics are connected to the generating functions through

$$i_0(t) \equiv \langle I_0(t)\rangle = \left. \frac{\partial F_0(s,t)}{\partial s} \right|_{s=1} \qquad (6.98)$$

and similarly for $i_1(t)$. Using Eqs. (6.88) and (6.89), we obtain

$$i_0(t) = 1 - G_0(t) + R_0 \int_0^t i_1(t - \tau) dG_0(\tau), \qquad (6.99)$$

$$i_1(t) = 1 - G_1(t) + R_1 \int_0^t i_1(t - \tau) dG_1(\tau). \qquad (6.100)$$

These calculations are straightforward extensions of those of the size of the giant component in percolation on static networks [Newman (2002)].

To know the asymptotic behaviour of $i_1(t)$, which also gives insights into asymptotic of $i_0(t)$ and $\langle S_0(t \to \infty) \rangle$, we use the value of α that satisfies

$$R_1 \int_0^\infty e^{-\alpha t} dG_1(t) = 1. \qquad (6.101)$$

If $R_1 > 1$, then α always exists and is positive. If $R_1 < 1$, then α is negative if it exists.

For a class of distributions $G_1(t)$ including the exponential distribution, corresponding to Poisson processes, the asymptotic number of new infected nodes is given by

$$i_1(t) \approx \frac{R_1 - 1}{\alpha R_1^2 \int_0^\infty t' e^{-\alpha t'} dG_1(t')} e^{\alpha t} \qquad (6.102)$$

with the value of α obtained from Eq. (6.101). For example, Poisson processes yield

$$G_1(t) = 1 - e^{-\lambda_1 t}, \qquad (6.103)$$

where λ_1 is the rate. Substitution of Eq. (6.103) in Eq. (6.101) yields

$$\alpha = \lambda_1(R_1 - 1). \qquad (6.104)$$

Substitution of Eqs. (6.103) and (6.104) in Eq. (6.102) yields

$$i_1(t) = e^{\alpha t} = e^{\lambda_1(R_1 - 1)t}. \qquad (6.105)$$

Therefore, $i_1(t)$ explodes or decays exponentially in t when $R_1 > 1$ or $R_1 < 1$, respectively. The exponential growth and decay indicated in Eqs. (6.102) and (6.105) extend the results for the Galton-Watson processes (Section 2.10) and the mean-field SI model (Eq. (6.59)).

For a class of $G_1(t)$ including long-tailed distributions, α does not exist for $R_1 < 1$. In this case, the long-time behaviour is given by

$$i_1(t) \approx \frac{1 - G_1(t)}{1 - R_1}. \qquad (6.106)$$

When $G_1(t)$ is a long-tailed distribution, $G_1(t)$ approaches unity slowly as t increases. Therefore, $i_1(t)$ decays to zero slowly with a long tail, explaining real data of computer viruses and viral information spreading. When $R_1 > 1$, Eq. (6.102) holds true with an appropriate $\alpha(> 0)$ value such that $i_1(t)$ grows exponentially.

6.4.5 *SIR dynamics on a tree-like stochastic temporal network*

The branching process model explained in the previous section was designed for understanding time evolution of SIR-type dynamics. In this model, the distribution of the number of secondary infections, $p_{i,r}$ automatically determines if the system is above or below the epidemic threshold. The epidemic threshold is independent of the distribution of the time to infection (i.e., $G_0(\tau)$ and $G_1(\tau)$). Empirically, the epidemic threshold may depend on temporal properties of nodes and links as well, which we explain in this section with the following model [Lambiotte *et al.* (2013)].

Let us consider the SIR process taking place on an infinite stochastic temporal network whose underlying topology is a tree of degree k (i.e., each node has k neighbours). On each link (i, j), events are independently generated according to the probability density function of inter-event times, $\psi_{ij}(\tau)$. As in the previous section, we initially infect a single node v_i at time 0. The probability distribution of time t until v_i contacts and infects v_j is given by $\psi_{ij}^{\rm w}(t)$ due to the waiting-time paradox. We also need the probability distribution function of the time to recovery, denoted by $p^{\rm I \to R}(t)$, to completely define the process. Empirically, $p^{\rm I \to R}(t)$ tends to have a shorter dispersion than the exponential distribution does and is often approximated by a gamma distribution whose CV < 1 [Lloyd (2001)] or more succinctly by the delta distribution.

Because v_i can infect v_j only if v_i remains infected at the time of the contact with v_j, the probability density function with which v_i infects v_j at time t is given by

$$p_{ij}^{\rm infect}(t) = \psi_{ij}^{\rm w}(t) \int_t^\infty p^{\rm I \to R}(t') dt'. \tag{6.107}$$

The overall probability that v_i infects v_j before v_i recovers, called the transmissibility [Newman (2002)], is given by

$$p_{ij}^{\rm infect} = \int_0^\infty p_{ij}^{\rm infect}(t) dt. \tag{6.108}$$

If the links have the same distribution of inter-event times and hence $\psi_{ij}^{\rm w}(t) = \psi^{\rm w}(t)$, the epidemic threshold is given by $(k - 1)p_{ij}^{\rm infect} = (k - 1) \int_0^\infty \psi^{\rm w}(t) \int_t^\infty p^{\rm I \to R}(t') dt' dt = 1$. Therefore, the epidemic threshold depends on both the distribution of the time to infection and that to recovery. In general, a process would be more viral with a higher transmissibility (e.g., above the epidemic threshold) when $\psi^{\rm w}(t)$ has a mass of probability on smaller t values. However, neither the average waiting time, its variance nor their combination completely determines the epidemic threshold. We have to instead look at the relative position of the two distributions, $\psi^{\mathrm{w}}(t)$ and $p^{\mathrm{I}\to\mathrm{R}}(t)$.

6.5   Synchronisation

A node in a network may represent a dynamical object such as cell, human or mechanical component. Supposing that a link represents a communication pathway between two dynamical elements, networks composed of dynamical nodes can synchronise under some conditions. How likely synchronisation is attained depends on network structure as well as the coupling strength, equivalent to link weight. Even if we condition on the same number of nodes and links, synchronisability depends on network structure. Conditions for synchronisation in coupled linear and non-linear dynamics in networks have been broadly studied [Arenas *et al.* (2008)].

For simplicity, consider linear diffusive dynamics on networks in continuous-time given by

$$\frac{\mathrm{d}\boldsymbol{x}}{\mathrm{d}t} = -L\boldsymbol{x}, \qquad (6.109)$$

where $\boldsymbol{x} = (x_1 \;\cdots\; x_N)^{\top}$, x_i is the dynamical state of the ith node and L is the Laplacian matrix given by Eq. (3.17). The strength of the coupling is incorporated into L if we allow L to be a weighted Laplacian matrix. When the network is undirected and connected, the eigenvalues of L satisfy $0 = \lambda_1 < \lambda_2 \leq \cdots \leq \lambda_N$ (Section 3.5). The equilibrium of the dynamics is revealed if we put the left-hand side of Eq. (6.109) to zero. Because $L\boldsymbol{x} = 0$, the equilibrium state is the zero eigenvector of the Laplacian, which is the synchronised state given by $\boldsymbol{x} = (1 \;\cdots\; 1)^{\top}$. The relaxation speed is governed by the spectral gap, i.e., λ_2. Synchronisation occurs rapidly if λ_2 is large. For example, a uniform and positive coupling strength between pairs of nodes linearly scales λ_2 such that a strong coupling enhances synchronisation.

In the context of Laplacian dynamics, synchronisation in temporal networks started to be studied in control theory [Olfati-Saber *et al.* (2007)]. Synchronisation is more often called consensus in this field. Motivated by dynamic networks that mobile agents generate through flocking, e.g., coordinating mobile sensor platforms and platoons of autonomous vehicles, various conditions for consensus (i.e., synchronisation) in dynamic networks (in our term, temporal networks) have been clarified. If we include studies of switching systems, whose example is dynamics on temporal networks in the snapshot representation, the history of research is even longer [Liberzon (2003)]. A majority of studies in this domain uses the snapshot representation more often than the event-based representation of temporal networks. We also do so in this section.

Linear switching dynamics (often called switching networks) in continuous time, induced by the Laplacian, are given by

$$\frac{\mathrm{d}\boldsymbol{x}}{\mathrm{d}t} = -L_i\boldsymbol{x} \quad (i\tau \leq t < (i+1)\tau), \tag{6.110}$$

where τ is the duration for which each snapshot Laplacian is applied, and L_i represents the ith Laplacian.

If each snapshot is Eulerian (i.e., in-degree and out-degree are equal at each node; undirected networks always satisfy this condition) and strongly connected, then synchronisation occurs, and the speed of the dynamics towards synchronisation is at least $\min_i \lambda_2(L_i)$ [Olfati-Saber and Murray (2004)]. This result is intuitive. This and many other theorems providing conditions for synchronisation require (strong) connectedness of each snapshot. In practice, this is a very strong condition. If the aggregation time window is very large, this condition would be met, but then a lot of temporal information contained in the original temporal network would be dropped.

We are primarily interested in the case in which a snapshot contains relatively few nodes and links. Fortunately, other theory has been developed to dismiss the connectedness condition. A relaxed condition for synchronisation is typically connectedness of networks obtained from aggregation of a finite or infinite number of snapshots; connectedness for each snapshot is no longer required (see [Olfati-Saber *et al.* (2007)] for a review). For example, the connectedness of the network aggregated over time interval $[t_i, t_{i+1})$ for each i ensures asymptotic synchronisation in discrete-time dynamics [Jadbabaie *et al.* (2003); Ren and Beard (2005)] (see [Moreau (2005)] for directed networks). We are allowed to select $\{t_i\}$. The theorems require that an infinite sequence of such time intervals can be taken and that the intervals are bounded. For undirected networks, the boundedness condition can be omitted [Moreau (2005)].

Other types of theorems guarantee synchronisation in switching networks of linear and non-linear systems when the network is blinking, i.e., when switching is very fast ($\tau \to 0$) [Porfiri *et al.* (2006); Stilwell *et al.* (2006); So *et al.* (2008); Hasler *et al.* (2013a)]. In this situation, we can apply the so-called time averaging technique to prove that the behaviour of the blinking system converges to that on the time-averaged (in our term, aggregate), static network as $\tau \to 0$ (reviewed in [Hasler *et al.* (2013a)]). In fact, synchronisation can also occur when τ is finite and smaller than a certain threshold [Hasler *et al.* (2013b)].

When each snapshot is not necessarily a connected network, most theoretical results have focused on the possibility of synchronisation. However, we are also interested in the relaxation time to synchronisation. Fujiwara and colleagues numerically studied a system of mobile and coupled nonlinear phase oscillators [Fujiwara *et al.* (2011)]. In their model, oscillators randomly move around in a two-dimensional space. Two oscillators interact if and only if they are located within distance d. The phases of individual oscillators are updated instantaneously with a period of τ, consistent with the snapshot representation of temporal networks. Instantaneous updating obeys

$$\theta_i(t+\tau) = \theta_i(t) + \sum_{j=1; d_{ij}<d}^{N} \sin\left[\theta_j(t) - \theta_i(t)\right], \qquad (6.111)$$

where $\theta_i(t)$ is the phase of the ith oscillator at time t, between 0 and 2π, and d_{ij} is the distance between oscillators i and j in the two-dimensional space. The sin term in Eq. (6.111) indicates that when the phases of the two oscillators are not too far, the oscillators attract each other. Note that $\sin\left[\theta_j(t) - \theta_i(t)\right] \approx \theta_j(t) - \theta_i(t)$ when $\theta_j(t)$ and $\theta_i(t)$ are close to each other. The coupling strength is normalised to unity in Eq. (6.111). This dynamical system is interpreted as temporal dynamics because, due to the random movement of each oscillator, the network whose link is defined by (i, j) such that $d_{ij} < d$ varies from snapshot to snapshot.

Synchronisation requires relatively many interaction events (as represented by Eq. (6.111)) if τ is large [Fujiwara *et al.* (2011)]. A small τ makes the temporal network close to the aggregate network, the extreme case of which is the blinking system (with $\tau \to 0$). Therefore, the result implies that temporal networks as represented by a relatively large τ slow down synchronisation processes.

Unless d is large, each snapshot would not be a connected network. Therefore, the mathematical results for the relaxation time constant [Olfati-Saber and Murray (2004)] cannot be applied to this case even if Eq. (6.111) is linearised. In this situation, the Laplacian eigenvalues for the linear dynamics on switching networks (Eq. (6.110)), denoted by λ_i^{tp} ($0 = \lambda_1^{\text{tp}} < \lambda_2^{\text{tp}} \le \cdots \le \lambda_N^{\text{tp}}$), can be related to that on the corresponding aggregate (therefore static) network, denoted by λ_i^{ag} ($0 = \lambda_1^{\text{ag}} < \lambda_2^{\text{ag}} \le \cdots \le \lambda_N^{\text{ap}}$).

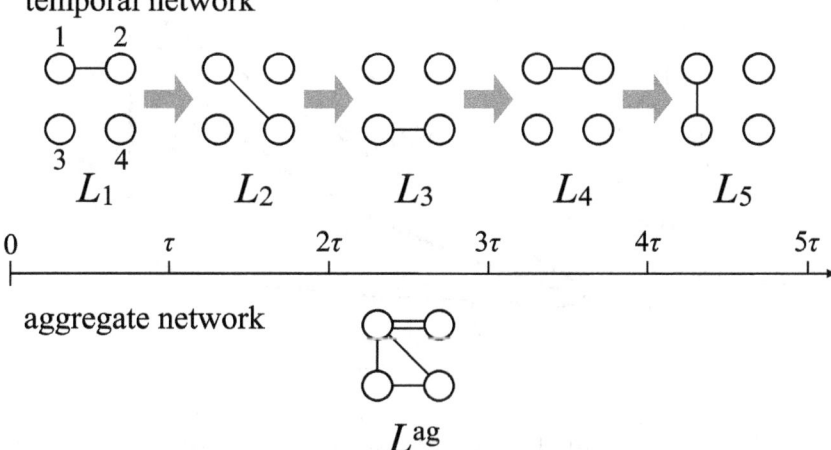

temporal network

aggregate network

L^{ag}

Fig. 6.7   Schematic of linear diffusive dynamics on a temporal network and the corresponding aggregate network when each snapshot contains a single event. $L^{\mathrm{ag}} = \sum_{i=1}^{r} L_i$ is the Laplacian of the aggregate network. In this example, the number of snapshots is equal to $r = 5$.

This relationship holds true if each snapshot contains only a single event, as shown in Fig. 6.7 [Masuda *et al.* (2013a)]. If we have r snapshots ($r = 5$ in Fig. 6.7), the dynamics on the temporal network should be compared with those on the aggregate network given by

$$\frac{\mathrm{d}\boldsymbol{x}}{\mathrm{d}t} = -L^* \boldsymbol{x} \quad (0 \le t \le r\tau), \tag{6.112}$$

where $L^* = \left(\sum_{i=1}^{r} L_i \right) / r \equiv L^{\mathrm{ag}}/r$ is the normalised Laplacian of the aggregate network; L^{ag} is the usual Laplacian of the aggregate network. The normalisation factor r is introduced to let each event be used for τ "weight × time" in both temporal and aggregate dynamics, which makes the comparison fair. For example, in Fig. 6.7, the event connecting nodes 3 and 4 appears in the third snapshot with a unity weight in the temporal network. In the aggregate network, the same event appears throughout the dynamics, i.e., for the duration of the five snapshots, with weight 1/5.

Under this and some additional conditions, the Laplacian eigenvalues of the two dynamics are related by

$$\lambda_i^{\mathrm{tp}} = -\frac{1}{\tau} \ln \left[1 - \frac{1 - \exp(-2\tau)}{2} \lambda_i^{\mathrm{ag}} \right] \quad (1 \le i \le N). \tag{6.113}$$

Equation (6.113) for $\tau = 0.1$, 1 and 10 is shown in Fig. 6.8. The figure indicates that λ_i^{tp} is always smaller (i.e., closer to zero) than λ_i^{ag}. Because

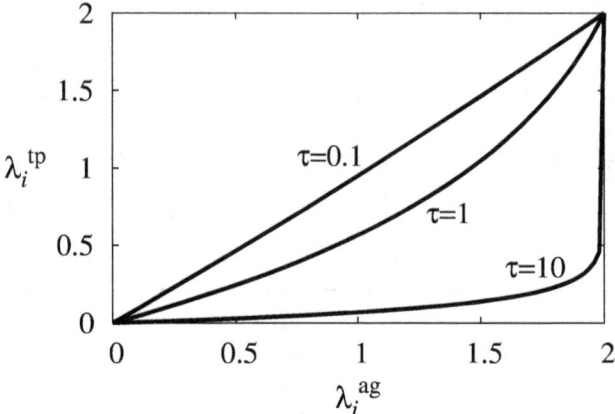

Fig. 6.8 Relationship between the Laplacian eigenvalue for the temporal dynamics, λ_i^{tp}, and that for the aggregate dynamics, λ_i^{ag}, which is given by Eq. (6.113).

a smaller Laplacian eigenvalue implies slower relaxation, temporal networks slow down synchronisation dynamics. The slowing-down effect is magnified as τ increases. When τ is small ($\tau = 0.1$ in Fig. 6.8), λ_i^{tp} and λ_i^{ag} are almost the same, corresponding to the blinking scheme.

Appendix A

Discrete-time random walks on the line

To understand the asymptotic behaviour of the random walk on the line in discrete time, it is useful to introduce the notion of stable distribution. A distribution $g(x)$ is said to be stable if it verifies the following invariance relation defined by its Fourier transform:

$$\hat{g}(a_1 k)\hat{g}(a_2 k) = \hat{g}(ak), \qquad (A.1)$$

where a_1 and a_2 are any positive constants and a is a function of a_1 and a_2. $g(x)$ is stable if $a_1 X_1 + a_2 X_2$ has the same distribution as $a X_3$, where X_1, X_2 and X_3 are independent random variables drawn from $g(x)$. In other words, a linear combination of two independent copies of the random variable has the same distribution, up to a scale parameter. Solutions of Eq. (A.1) include the Lévy distribution defined in the Fourier domain as

$$\hat{g}(k) = e^{-C|k|^\beta}, \qquad (A.2)$$

where C is a positive constant and $0 < \beta \leq 2$. For this distribution, we obtain $a = \left(a_1^\beta + a_2^\beta\right)^{1/\beta}$.

If the structure function of the random walk on the line is a Lévy distribution, i.e., $\hat{f}(k) = e^{-C|k|^\beta}$, function

$$\hat{p}(k;t) = \left[\hat{f}(k)\right]^t = e^{-Ct|k|^\beta} = \hat{f}(t^{1/\beta}k) \qquad (A.3)$$

is also Lévy, with the same value of β, for any $t > 0$. Equation (A.3) respects self-similarity such that the shape of the distribution is preserved up to scaling. The effect of time is simply to multiply the argument of the function.

The Lévy distribution with $\beta = 2$ is the Gaussian distribution because

Eq. (2.63), i.e., the inverse Fourier transform of Eq. (A.3), yields

$$p(x;t) = \frac{1}{2\pi} \int_{-\infty}^{\infty} e^{-Ctk^2} e^{ikx} \mathrm{d}k$$

$$= \frac{1}{(4\pi Ct)^{1/2}} e^{-\frac{x^2}{4Ct}} \tag{A.4}$$

with $\langle x \rangle = 0$ and $\langle x^2 \rangle = 2Ct$. If the structure function is Gaussian, we thus observe diffusive behaviour where the position of the random walk obeys the Gaussian distribution and the variance of the position grows linearly with time.

Lévy stable distributions play a central role in the theory of random walks for the following reason. Any structure function, which is not necessarily the Lévy stable distribution, leads to a solution approaching Eq. (A.3) in the Fourier domain. If the first two moments of the structure function are finite, we obtain

$$\hat{f}(k) = \langle e^{-ikr} \rangle = e^{-ik\langle r \rangle - \frac{1}{2}k^2 \langle (r - \langle r \rangle)^2 \rangle + \cdots}, \tag{A.5}$$

for small k, where r is the displacement in a single jump. Then, the solution of the random walk is given by

$$p(x;t) = \frac{1}{2\pi} \int_{-\infty}^{\infty} e^{ikx} e^{-ikt\langle r \rangle - \frac{1}{2}k^2 t \langle (r - \langle r \rangle)^2 \rangle + \cdots} \mathrm{d}k. \tag{A.6}$$

In terms of the drift constant $v \equiv \langle r \rangle$ and the diffusion constant $D \equiv \langle (r - \langle r \rangle)^2 \rangle / 2$, Eq. (A.6) is rewritten as

$$p(x;t) = \frac{1}{2\pi} \int_{-\infty}^{\infty} e^{ik(x-vt)} e^{-k^2 Dt} \mathrm{d}k$$

$$= \frac{1}{(2\pi Dt)^{1/2}} e^{-\frac{(x-vt)^2}{4Dt}}. \tag{A.7}$$

Therefore, any structure function whose first two moments are finite asymptotically leads to a Gaussian profile (Eq. (2.64)). This result is the central limit theorem for the summation of the jump sizes, which are independent random variables.

When the second moment of the structure function is infinite, Eq. (A.5) is invalid. In this case, an expansion of $\hat{f}(k) = e^{-C|k|^\beta}$ ($0 < \beta < 2$) yields

$$\hat{f}(k) = 1 - C|k|^\beta. \tag{A.8}$$

By applying the Tauberian theorem (Section 2.4) to Eq. (A.8), we obtain a power-law tail in the physical space given by $f(r) \approx |r|^{-\beta-1}$. In the random walk driven by Eq. (A.8), the distribution of the walker's position converges to a stable Lévy distribution (Eq. (A.3)) with the same value of β as $t \to \infty$. The random walk is super-diffusive, i.e., the variance of the position grows faster than linearly with time [Klafter and Sokolov (2011)].

Appendix B

Transient and absorbing states of Markov chains

Let us consider a Markov chain composed of N_1 transient states and N_2 absorbing states, where $N_1 + N_2 = N$. We rearrange the label of the states, such that states $1, \ldots, N_1$ are transient states and states $N_1 + 1, \ldots, N$ are absorbing states. Then, the transition matrix takes the form

$$T = \begin{pmatrix} Q & R \\ 0 & I \end{pmatrix}, \tag{B.1}$$

where Q is an $N_1 \times N_1$ square matrix, describing transitions between transient states; R is an $N_1 \times N_2$, matrix describing transitions from transient states to absorbing states; I is the $N_2 \times N_2$ identity matrix associated with the absorbing states. This ordering of the states has the advantage that powers of the transition matrix are expressed as products of the submatrices as follows:

$$T^t = \begin{pmatrix} Q^t & R + QR + \cdots + QR^{t-1} \\ 0 & I \end{pmatrix}. \tag{B.2}$$

Let us first calculate the average number of visits to transient state j, starting from transient state i, before an absorbing state is reached. This number is given by the (i, j) element of

$$W = \sum_{t=0}^{\infty} Q^t. \tag{B.3}$$

The (i, j) element of each term on the right-hand side of Eq. (B.3) represents the probability that state j is visited at time t, starting from state i. Equation (B.3) is written as

$$W = (I - Q)^{-1}. \tag{B.4}$$

Matrix W is called the fundamental matrix associated with Q.

The structure of matrix T also enables us to calculate the probability U_{ij} that the process ends in absorbing state j when it starts from transient state i. This is called the exit probability. The probability of arriving at state j exactly after t steps is given by the (i, j) element of $Q^{t-1}R$. Therefore, we obtain

$$U = \sum_{t=1}^{\infty} Q^{t-1}R = WR. \tag{B.5}$$

Appendix C

Derivation of the degree distribution of the Barabási-Albert model

The degree distribution of the BA model can be derived in two relatively easy manners. In the first method, we approximate k and t, which are integer variables, to be continuous variables and also discard stochasticity in the model [Barabási and Albert (1999); Barabási et $al.$ (1999)]. To carry out the analysis, we start with the observation that a generated network gains one node per unit time. The initial condition is given by m_0 nodes at $t = 0$. For simplicity, we set $m = m_0$ and assume that the initial network is the complete graph such that it has $m(m-1)/2$ links.

Consider what happens at time $t = N - m$, when there are N nodes. Because one node carries m links, there are $m(m-1)/2 + m(t-1)$ links in total at time $t - 1$. Using the fact that one link gives a contribution of two to the sum of the degrees over the network and that there are $N - 1$ nodes at time $t - 1 = N - 1 - m$, we obtain

$$\sum_{j=1}^{N-1} k_j = 2 \times (\text{number of links at } t = N - 1 - m)$$

$$= 2 \left[\frac{m(m-1)}{2} + mt \right] \approx 2mt \qquad (C.1)$$

for large N.

The degree of each node grows on average according to

$$\frac{\mathrm{d}k_i}{\mathrm{d}t} \approx m\Pi(k_i) = \frac{k_i}{2t}. \qquad (C.2)$$

By solving Eq. (C.2) under the initial condition $k_i(t_i) = m$, we obtain

$$k_i(t) = m \left(\frac{t}{t_i} \right)^{\frac{1}{2}}. \qquad (C.3)$$

Equation (C.3) states that an older node (with smaller t_i) has a large degree on average. Using this monotonic relationship between k_i and t_i, we obtain

$$p\left[k_i(t) < k\right] = p\left[t_i > \frac{m^2 t}{k^2}\right] = \frac{1}{m+t}\left(t - \frac{m^2 t}{k^2}\right). \qquad (C.4)$$

The first equality in Eq. (C.4) is due to Eq. (C.3). The second equality holds true because, among the $m+t$ nodes at time t, we are interested in the nodes satisfying $t_i > m^2 t/k^2$, i.e., $t_i = (m^2 t/k^2) + 1, (m^2 t/k^2) + 2, \ldots,$ t; there are $t - (m^2 t/k^2)$ such nodes.

By differentiating Eq. (C.4) with respect to k, we obtain

$$p(k) = \frac{\partial p\left[k_i(t) < k\right]}{\partial k} = \frac{2m^2 t}{k^3 (m+t)} \propto k^{-3}. \qquad (C.5)$$

Another more exact approach to the degree distribution of the BA model is via master equations. In this approach, k and t are not approximated to be continuous or is connected by a deterministic relationship such as Eq. (C.3). Denote by $p(k, t_i, t)$ the probability that a node v_i that has joined at time t_i has degree k at time t. The master equation for $p(k, t_i, t)$ is given by

$$p(k, t_i, t+1) = \frac{k-1}{2t} p(k-1, t_i, t) + \left(1 - \frac{k}{2t}\right) p(k, t_i, t) \qquad (C.6)$$

because k increases by one with probability $m\Pi(k) \approx k/2t$ and does not change with probability $1 - k/2t$ in a unit time. When N is large, we obtain

$$p(k) = \lim_{t \to \infty} \frac{\sum_{t_i} p(k, t_i, t)}{t}. \qquad (C.7)$$

The normalisation factor $1/t$ comes from the fact that there are $t + m \approx t$ nodes at time t.

The node that joins at time $t_i = t + 1$ has been absent at time t, such that $p(k, t+1, t) = 0$. By using this and summing Eq. (C.6) over t_i, we obtain

$$\sum_{t_i=1}^{t+1} p(k, t_i, t+1) = \frac{k-1}{2t} \sum_{t_i=1}^{t} p(k-1, t_i, t) + \left(1 - \frac{k}{2t}\right) \sum_{t_i=1}^{t} p(k, t_i, t). \quad (C.8)$$

By substituting $p(k) \approx \sum_{t_i=1}^{t} p(k, t_i, t)/t = \sum_{t_i=1}^{t+1} p(k, t_i, t+1)/(t+1)$ and $p(k-1) \approx \sum_{t_i=1}^{t} p(k-1, t_i, t)/t$ in Eq. (C.8), we obtain

$$(t+1)p(k) = \frac{k-1}{2t} tp(k-1) + \left(1 - \frac{k}{2t}\right) tp(k). \qquad (C.9)$$

Equation (C.9) yields

$$p(k) = \frac{k-1}{k+2} p(k-1) \quad (k \ge m+1), \qquad (C.10)$$

which yields

$$p(k) \propto \frac{1}{k(k+1)(k+2)} \propto k^{-3}. \qquad (C.11)$$

Bibliography

Adamic, L. A. and Adar, E. (2003). Friends and neighbors on the Web, *Social Networks* **25**, pp. 211–230.

Aggarwal, C. and Subbian, K. (2014). Evolutionary network analysis: A survey, *ACM Computing Surveys* **47**, article No. 10.

Ahn, Y. -Y., Bagrow, J. P. and Lehmann, S. (2010). Link communities reveal multiscale complexity in networks, *Nature* **466**, pp. 761–764.

Akoglu, L., Tong, H. and Koutra, D. (2015). Graph based anomaly detection and description: A survey, *Data Mining and Knowledge Discovery* **29**, pp. 626–688.

Allegrini, P., Menicucci, D., Bedini, R., Fronzoni, L., Gemignani, A., Grigolini, P., West, B. J. and Paradisi, P. (2009). Spontaneous brain activity as a source of ideal $1/f$ noise, *Physical Review E* **80**, article No. 061914.

Allen, A. O. (1990). *Probability, Statistics, and Queueing Theory: With Computer Science Applications, Second Edition* (Academic Press, Boston).

Anderson, R. M. and May, R. M. (1991). *Infectious Diseases of Humans — Dynamics and Control* (Oxford University Press, Oxford).

Anteneodo, C., Malmgren, R. D. and Chialvo, D. R. (2010). Poissonian bursts in e-mail correspondence, *European Physical Journal B* **75**, pp. 389–394.

Arenas, A., Díaz-Guilera, A., Kurths, J., Moreno, Y. and Zhou, C. (2008). Synchronization in complex networks, *Physics Reports* **469**, pp. 93–153.

Backlund, V. -P., Saramäki, J. and Pan R. K. (2014). Effects of temporal correlations on cascades: Threshold models on temporal networks. *Physical Review E* **89**, article No. 062815.

Backstrom, L., Boldi, P., Rosa, M., Ugander, J. and Vigna, S. (2012). Four degrees of separation, *Proceedings of the 4th Annual ACM Web Science Conference (WebSci)*, pp. 33–42.

Balcan, D. and Vespignani, A. (2011). Phase transitions in contagion processes mediated by recurrent mobility patterns, *Nature Physics* **7**, pp. 581–586.

Barabási, A. -L. (2005). The origin of bursts and heavy tails in human dynamics, *Nature* **435**, pp. 207–211.

Barabási, A. -L. (2016). *Network Science* (Cambridge University Press, Cambridge).

Barabási, A. -L. and Albert, R. (1999). Emergence of scaling in random networks, *Science* **286**, pp. 509–512.

Barabási, A. -L., Albert, R. and Jeong, H. (1999). Mean-field theory for scale-free random networks, *Physica A* **272**, pp. 173–187.

Barrat, A., Barthélemy, M., Pastor-Satorras, R. and Vespignani, A. (2004). The architecture of complex weighted networks, *Proceedings of the National Academy of Sciences of the United States of America* **101**, pp. 3747–3752.

Barrat, A. and Pastor-Satorras, R. (2005). Rate equation approach for correlations in growing network models, *Physical Review E* **71**, article No. 036127.

Barthélemy, M. (2011). Spatial networks, *Physics Reports* **499**, pp. 1–101.

Bassett, D. S., Porter, M. A., Wymbs, N. F., Grafton, S. T., Carlson, J. M. and Mucha, P. J. (2013). Robust detection of dynamic community structure in networks, *Chaos* **23**, article No. 013142.

Bazzi, M., Porter, M. A., Williams, S., McDonald, M., Fenn, D. J. and Howison, S. D. (2016). Community detection in temporal multilayer networks, with an application to correlation networks, *Multiscale Modeling and Simulation* **1**, pp. 1–41.

Belik, V., Geisel, T. and Brockmann, D. (2011). Natural human mobility patterns and spatial spread of infectious diseases, *Physical Review X* **1**, article No. 011001.

Berman, K. A. (1996). Vulnerability of scheduled networks and a generalization of Menger's theorem, *Networks* **28**, pp. 125–134.

Bhadra, S. and Ferreira, A. (2003). Complexity of connected components in evolving graphs and the computation of multicast trees in dynamic networks, *Lecture Notes in Computer Science* **2865**, pp. 259–270.

Blondel, V. D., Guillaume, J. -L., Lambiotte, R. and Lefebvre, E. (2008). Fast unfolding of communities in large networks, *Journal of Statistical Mechanics*, article No. P10008.

Boguñá, M., Lafuerza, L. F., Toral, R. and Serrano, M. Á. (2014). Simulating non-Markovian stochastic processes, *Physical Review E* **90**, article No. 042108.

Boldi, P., Santini, M. and Vigna, S. (2005). PageRank as a function of the damping factor, *Proceedings of the 14th International Conference on World Wide Web (WWW)*, pp. 557–566.

Bollobás, B. (2001). *Random Graphs, Second Edition* (Cambridge University Press, Cambridge).

Bollobás, B. and Riordan, O. (2004). The diameter of a scale-free random graph, *Combinatorica* **24**, pp. 5–34.

Bornholdt, S. and Schuster, H. G. (eds.) (2003). *Handbook of Graphs and Networks — From the Genome to the Internet* (Wiley-VCH, Weinheim).

Brandes, U., Delling, D., Gaertler, M., Görke, R., Hoefer, M., Nikoloski, Z. and Wagner, D. (2008). On modularity clustering, *IEEE Transactions on Knowledge and Data Engineering* **20**, pp. 172–188.

Brin, S. and Page, L. (1998). Anatomy of a large-scale hypertextual web search engine, *Proceedings of the Seventh International World Wide Web Conference (WWW)*, pp. 107–117.

Brinkmeier, M. (2006). PageRank revisited, *ACM Transactions on Internet Technology* **6**, pp. 282–301.

Buldyrev, S. V. (2010). Fractals in biology, In Meyers, R. A. (ed.), *Encyclopedia of Complexity and Systems Science* (Springer Science + Business Media, New York), pp. 3779–3802.

Caldarelli, G., Capocci, A., De Los Rios, P. and Muñoz, M. A. (2002). Scale-free networks from varying vertex intrinsic fitness, *Physical Review Letters* **89**, article No. 258702.

Casteigts, A., Flocchini, P., Quattrociocchi, W. and Santoro, N. (2011). Time-varying graphs and dynamic networks, *Lecture Notes in Computer Science* **6811**, pp. 346–359.

Cattuto, C., Van den Broeck, W., Barrat, A., Colizza, V., Pinton, J. -F. and Vespignani, A. (2010). Dynamics of person-to-person interactions from distributed RFID sensor networks, *PLOS ONE* **5**, article No. e11596.

Centola, D. and Macy, M. (2007). Complex contagions and the weakness of long ties, *American Journal of Sociology* **113**, pp. 702–734.

Chandola, V., Banerjee, A. and Kumar, V. (2009). Anomaly detection: A survey, *ACM Computing Surveys* **41**, article No. 15.

Chen, Z., Vijayan, S., Barbieri, R., Wilson, M. A. and Brown, E. N. (2009). Discrete- and continuous-time probabilistic models and algorithms for inferring neuronal UP and DOWN states, *Neural Computation* **21**, pp. 1797–1862.

Cheng, E., Grossman, J. W. and Lipman, M. J. (2003). Time-stamped graphs and their associated influence digraphs, *Discrete Applied Mathematics* **128**, pp. 317–335.

Chung, F. R. K. (1997). *Spectral Graph Theory* (American Mathematical Society, Providence).

Clauset, A. (2010). Inference, models and simulation for complex systems, `http://tuvalu.santafe.edu/~aaronc/courses/7000/`

Clauset, A. and Eagle, N. (2007). Persistence and periodicity in a dynamic proximity network, *Proceedings of the DIMACS Workshop on Computational Methods for Dynamic Interaction Networks*, available as Preprint arXiv:1211.7343v1.

Clauset, A., Shalizi, C. R. and Newman, M. E. J. (2009). Power-law distributions in empirical data, *SIAM Review* **51**, pp. 661–703.

Clementi, A. E. F., Macci, C., Monti, A., Pasquale, F. and Silvestri, R. (2008). Flooding time in edge-Markovian dynamic graphs, *Proceedings of the Twenty-seventh ACM Symposium on Principles of Distributed Computing (PODC)*, pp. 213–222.

Clementi, A. E. F., Macci, C., Monti, A., Pasquale, F. and Silvestri, R. (2010). Flooding time of edge-Markovian evolving graphs, *SIAM Journal of Discrete Mathematics* **24**, pp. 1694–1712.

Cobham, A. (1954). Priority assignment in waiting time problems, *Journal of the Operations Research Society of America* **2**, pp. 70–76.

Cohen, R. and Havlin, S. (2003). Scale-free networks are ultrasmall, *Physical Review Letters* **90**, article No. 058701.

Colizza, V., Barrat, A., Barthélemy, M. and Vespignani, A. (2006). The role of the airline transportation network in the prediction and predictability of global epidemics, *Proceedings of the National Academy of Sciences of the United States of America* **103**, pp. 2015–2020.

Colizza, V., Pastor-Satorras, R. and Vespignani, A. (2007). Reaction-diffusion processes and metapopulation models in heterogeneous networks, *Nature Physics* **3**, pp. 276–282.

Colizza, V. and Vespignani, A. (2008). Epidemic modeling in metapopulation systems with heterogeneous coupling pattern: Theory and simulations, *Journal of Theoretical Biology* **251**, pp. 450–467.

Cox, D. R. (1962). *Renewal Theory* (Methuen & Co. Ltd, Frome).

Cvetković, D., Rowlinson, P. and Simić, S. (2010). *An Introduction to the Theory of Graph Spectra* (Cambridge University Press, Cambridge).

de Solla Price, D. (1976). A general theory of bibliometric and other cumulative advantage processes, *Journal of the American Society for Information Science* **27**, pp. 292–306.

Delvenne, J. -C., Lambiotte, R. and Rocha, L. E. C. (2015). Diffusion on networked systems is a question of time or structure, *Nature Communications* **6**, article No. 7366.

Delvenne, J. -C., Schaub, M. T., Yaliraki, S. N. and Barahona, M. (2013). The stability of a graph partition: A dynamics-based framework for community detection, In: Mukherjee, A., Choudhury, M., Peruani, F., Ganguly, N. and Mitra, B. (eds.), *Dynamics on and of Complex Networks, Volume 2 — Applications to Time-varying Dynamical Systems* (Springer, New York), pp. 221–242.

Diekmann, O. and Heesterbeek, J. A. P. (2000). *Mathematical Epidemiology of Infectious Diseases — Model Building, Analysis and Interpretation* (John Wiley & Sons, Ltd., Chichester).

Dodds, P. S. and Watts, D. J. (2004). Universal behavior in a generalized model of contagion, *Physical Review Letters* **92**, article No. 218701.

Eckmann, J. -P., Moses, E. and Sergi, D. (2004). Entropy of dialogues creates coherent structures in e-mail traffic, *Proceedings of the National Academy of Sciences of the United States of America* **101**, pp. 14333–14337.

Escola, S., Fontanini, A., Katz, D. and Paninski, L. (2011). Hidden Markov models for the stimulus-response relationships of multistate neural systems, *Neural Computation* **23**, pp. 1071–1132.

Evans, M. R. (2000). Phase transitions in one-dimensional nonequilibrium systems, *Brazilian Journal of Physics* **30**, pp. 42–57.

Evans, T. S. and Lambiotte, R. (2009). Line graphs, link partitions and overlapping communities, *Physical Review E* **80**, article No. 016105.

Feld, S. L. (1991). Why your friends have more friends than you do, *American Journal of Sociology* **96**, pp. 1464–1477.

Feller, W. (1949). Fluctuation theory of recurrent events, *Transactions of the American Mathematical Society* **67**, pp. 98–119.

Feller, W. (1971). *An Introduction to Probability Theory and Its Applications, Volume II, Second Edition* (John Wiley & Sons).

Fernández-Gracia, J., Eguíluz, V. M. and San Miguel, M. (2011). Update rules and interevent time distributions: Slow ordering versus no ordering in the voter model, *Physical Review E* **84**, article No. 015103(R).

Fortunato, S. (2010). Community detection in graphs, *Physics Reports* **486**, pp. 75–174.

Fortunato, S. and Barthélemy, M. (2007). Resolution limit in community detection, *Proceedings of the National Academy of Sciences of the United States of America* **104**, pp. 36–41.

Fortunato, S., Boguñá, M., Flammini, A. and Menczer, F. (2008). Approximating PageRank from in-degree, *Lecture Notes in Computer Science* **4936**, pp. 59–71.

Friggeri, A., Chelius, G. and Fleury, E. (2011). Triangles to capture social cohesion, *Proceedings of IEEE Third International Conference on Social Computing (SocialCom)*, pp. 258–265.

Fujiwara, N., Kurths, J. and Díaz-Guilera, A. (2011). Synchronization in networks of mobile oscillators, *Physical Review E* **83**, article No. 025101(R).

Gauvin, L., Panisson, A. and Cattuto, C. (2014). Detecting the community structure and activity patterns of temporal networks: A non-negative tensor factorization approach, *PLOS ONE* **9**, article No. e86028.

Gillespie, D. T. (1976). A general method for numerically simulating the stochastic time evolution of coupled chemical reactions, *Journal of Computational Physics* **22**, pp. 403–434.

Gillespie, D. T. (1977). Exact stochastic simulation of coupled chemical reactions, *Journal of Physical Chemistry* **81**, pp. 2340–2361.

Goh, K. -I. and Barabási, A. -L. (2008). Burstiness and memory in complex systems, *EPL* **81**, article No. 48002.

Goh, K. -I., Kahng, B. and Kim, D. (2001). Universal behavior of load distribution in scale-free networks, *Physics Review Letters* **87**, article No. 278701.

Good, B. H., de Montjoye, Y. -A. and Clauset, A. (2010). Performance of modularity maximization in practical contexts, *Physical Review E* **81**, article No. 046106.

Granger, C. W. J. (1969). Investigating causal relations by econometric models and cross-spectral methods, *Econometrica* **37**, pp. 424–438.

Granovetter, M. (1978). Threshold models of collective behavior, *American Journal of Sociology* **83**, pp. 1420–1443.

Granovetter, M. S. (1973). The strength of weak ties, *American Journal of Sociology* **78**, pp. 1360–1380.

Greene, D., Doyle, D. and Cunningham, P. (2010). Tracking the evolution of communities in dynamic social networks, *Proceedings of 2010 International Conference on Advances in Social Networks Analysis and Mining (ASONAM)*, pp. 176–183.

Grigolini, P., Palatella, L. and Raffaelli, G. (2001). Asymmetric anomalous diffusion: An efficient way to detect memory in time series, *Fractals* **9**, pp. 439–449.

Grindrod, P. and Higham, D. J. (2013). A matrix iteration for dynamic network summaries, *SIAM Review* **55**, pp. 118–128.

Grindrod, P. and Higham, D. J. (2014). A dynamical systems view of network centrality, *Proceedings of the Royal Society A* **470**, article No. 20130835.

Grindrod, P., Parsons, M. C., Higham, D. J. and Estrada, E. (2011). Communicability across evolving networks, *Physical Review E* **83**, article No. 046120.

Grinstein, G. and Linsker, R. (2006). Biased diffusion and universality in model queues, *Physical Review Letters* **97**, article No. 130201.

Grinstein, G. and Linsker, R. (2008). Power-law and exponential tails in a stochastic priority-based model queue, *Physical Review E* **77**, article No. 012101.

Gross, T. and Blasius, B. (2008). Adaptive coevolutionary networks: A review, *Journal of the Royal Society Interface* **5**, pp. 259–271.

Gross, T. and Sayama, H. (eds.) (2009). *Adaptive Networks — Theory, Models and Applications* (Springer, Dortrecht).

Guimerà, R., Sales-Pardo, M. and Amaral, L. A. N. (2004). Modularity from fluctuations in random graphs and complex networks, *Physical Review E* **70**, article No. 025101(R).

Hanneke, S., Fu, W. and Xing, E. P. (2010). Discrete temporal models of social networks, *Electronic Journal of Statistics* **4**, pp. 585–605.

Hanneke, S. and Xing, E. P. (2007). Discrete temporal models of social networks, *Lecture Notes in Computer Science* **4503**, pp.115–125.

Hanski, I. (1998). Metapopulation dynamics, *Nature* **396**, pp. 41–49.

Hasan, M. A. and Zaki, M. J. (2011). A survey of link prediction in social networks, In: Aggarwal, C. C. (ed.), *Social Network Data Analytics* (Springer, New York), pp. 243–275.

Hasler, M., Belykh, V. and Belykh, I. (2013a). Dynamics of stochastically blinking systems. Part I: Finite time properties, *SIAM Journal of Applied Dynamical Systems* **12**, pp. 1007–1030.

Hasler, M., Belykh, V. and Belykh, I. (2013b). Dynamics of stochastically blinking systems. Part II: Asymptotic properties, *SIAM Journal of Applied Dynamical Systems* **12**, pp. 1031–1084.

Hawkes, A. G. (1971a). Point spectra of some mutually exciting point processes, *Journal of the Royal Statistical Society Series B* **33**, pp. 438–443.

Hawkes, A. G. (1971b). Spectra of some self-exciting and mutually exciting point processes, *Biometrika* **58**, pp. 83–90.

Hawkes, A. G. and Oakes, D. (1974). A cluster process representation of a self-exciting process, *Journal of Applied Probability* **11**, pp. 493–503.

Helmstetter, A. and Sornette, D. (2002). Subcritical and supercritical regimes in epidemic models of earthquake aftershocks, *Journal of Geophysical Research* **107**, article No. 2237.

Hoffmann, T., Porter, M. A. and Lambiotte, R. (2012). Generalized master equations for non-Poisson dynamics on networks, *Physical Review E* **86**, article No. 046102.

Holland, P. W. and Leinhardt, S. (1977). A dynamic model for social networks, *Journal of Mathematical Sociology* **5**, pp. 5–20.

Holme, P. (2005). Network reachability of real-world contact sequences, *Physical Review E* **71**, article No. 046119.

Holme, P. (2015). Modern temporal network theory: A colloquium, *European*

Physical Journal B **88**, article No. 234.

Holme, P. and Saramäki, J. (2012). Temporal networks, *Physics Reports* **519**, pp. 97–125.

Holme, P. and Saramäki, J. (eds.) (2013). *Temporal Networks* (Springer-Verlag, Berlin).

Hu, K., Ivanov, P. C., Chen, Z., Carpena, P. and Stanley, H. E. (2001). Effect of trends on detrended fluctuation analysis, *Physical Review E* **64**, article No. 011114.

Hufnagel, L., Brockmann, D. and Geisel, T. (2004). Forecast and control of epidemics in a globalized world, *Proceedings of the National Academy of Sciences of the United States of America* **101**, pp. 15124–15129.

Iribarren, J. L. and Moro, E. (2009). Impact of human activity patterns on the dynamics of information diffusion, *Physical Review Letters* **103**, article No. 038702.

Iribarren, J. L. and Moro, E. (2011). Branching dynamics of viral information spreading, *Physical Review E* **84**, article No. 046116.

Jadbabaie, A., Lin, J. and Morse, A. S. (2003). Coordination of groups of mobile autonomous agents using nearest neighbor rules, *IEEE Transactions on Automatic Control* **48**, pp. 988–1001.

Jo, H. -H., Karsai, M., Kertész, J. and Kaski, K. (2012). Circadian pattern and burstiness in mobile phone communication, *New Journal of Physics* **14**, article No. 013055.

Jo, H. -H., Perotti, J. I., Kaski, K. and Kertész, J. (2014). Analytically solvable model of spreading dynamics with non-Poissonian processes, *Physical Review X* **4**, article No. 011041.

Kagan, Y. Y. and Knopoff, L. (1981). Stochastic synthesis of earthquake catalogs, *Journal of Geophysical Research* **86**, pp. 2853–2862.

Kagan, Y. Y. and Knopoff, L. (1987). Statistical short-term earthquake prediction, *Science* **236**, pp. 1563–1567.

Kantelhardt, J. W., Koscielny-Bunde, E., Rego, H. H. A., Havlin, S. and Bunde A. (2001). Detecting long-range correlations with detrended fluctuation analysis, *Physica A* **295**, pp. 441–454.

Karimi, F. and Holme, P. (2013). Threshold model of cascades in empirical temporal networks, *Physica A* **392**, pp. 3476–3483.

Karrer, B. and Newman, M. E. J. (2010). Message passing approach for general epidemic models, *Physical Review E* **82**, article No. 016101.

Karsai, M., Kaski, K., Barabási, A. -L. and Kertész, J. (2012). Universal features of correlated bursty behaviour, *Scientific Reports* **2**, article No. 397.

Karsai, M., Perra, N. and Vespignani, A. (2014). Time varying networks and the weakness of strong ties, *Scientific Reports* **4**, article No. 4001.

Katz, L. (1953). A new status index derived from sociometric analysis, *Psychometrika* **18**, pp. 39–43.

Kawadia, V. and Sreenivasan, S. (2012). Sequential detection of temporal communities by estrangement confinement, *Scientific Reports* **2**, article No. 794.

Kempe, D., Kleinberg, J. and Kumar, A. (2000). Connectivity and inference problems for temporal networks, *Proceedings of the Thirty-second Annual*

ACM Symposium on Theory of Computing (STOC), pp. 504–513.

Kendall, D. G. (1950). An artificial realization of a simple "birth-and-death" process, *Journal of the Royal Statistical Society Series B* **12**, pp. 116–119.

Klafter, J. and Sokolov, I. M. (2011). *First Steps in Random Walks — From Tools to Applications* (Oxford University Press, Oxford).

Kleinberg, J. (2007). Cascading behavior in networks: Algorithmic and economic issues, In: Nisan, N., Roughgarden, T., Tardos, É. and Vazirani, V. V. (eds.), *Algorithmic Game Theory* (Cambridge University Press, Cambridge), pp. 613–632.

Klemm, K. and Eguíluz, V. M. (2002). Growing scale-free networks with small-world behavior, *Physical Review E* **65**, article No. 057102.

Kolda, T. G. and Bader, B. W. (2009). Tensor decompositions and applications, *SIAM Review* **51**, pp. 455–500.

Kossinets, G., Kleinberg, J. and Watts, D. (2008). The structure of information pathways in a social communication network, *Proceedings of the 14th ACM SIGKDD International Conference on Knowledge Discovery and Data Mining (KDD)*, pp. 435–443.

Kossinets, G. and Watts, D. J. (2006). Empirical analysis of an evolving social network, *Science* **311**, pp. 88–90.

Kovanen, L., Karsai, M., Kaski, K., Kertész, J. and Saramäki, J. (2011). Temporal motifs in time-dependent networks, *Journal of Statistical Mechanics*, article No. P11005.

Kovanen, L., Karsai, M., Kaski, K., Kertész, J. and Saramäki, J. (2013a). Temporal motifs, In: Holme, P. and Saramäki, J. (eds.), *Temporal Networks* (Springer, Berlin), pp. 119–133.

Kovanen, L., Kaski, K., Kertész, J. and Saramäki, J. (2013b). Temporal motifs reveal homophily, gender-specific patterns, and group talk in call sequences, *Proceedings of the National Academy of Sciences of the United States of America* **110**, pp. 18070–18075.

Krackhardt, D. and Handcock, M. S. (2007). Heider vs Simmel: Emergent features in dynamic structures, *Lecture Notes in Computer Science* **4503**, pp. 14–27.

Lambiotte, R., Delvenne, J. -C. and Barahona, M. (2009). Laplacian dynamics and multiscale modular structure in networks, Preprint: arXiv:0812.1770v2.

Lambiotte, R. and Rosvall, M. (2012). Ranking and clustering of nodes in networks with smart teleportation, *Physical Review E* **85**, article No. 056107.

Lambiotte, R., Salnikov, V. and Rosvall, M. (2015). Effect of memory on the dynamics of random walks on networks, *Journal of Complex Networks* **3**, pp. 177–188.

Lambiotte, R., Tabourier, L. and Delvenne, J. -C (2013). Burstiness and spreading on temporal networks, *European Physical Journal B* **86**, article No. 320.

Lamport, L. (1978). Time, clocks, and the ordering of events in a distributed system, *Communications of the ACM* **21**, pp. 558–565.

Lancichinetti, A., Radicchi, F., Ramasco, J. J. and Fortunato, S. (2011). Finding statistically significant communities in networks, *PLOS ONE* **6**, article No. e18961.

Langville, A. N. and Meyer, C. D. (2004). Deeper inside PageRank, *Internet*

Mathematics **1**, pp. 335–380.

Langville, A. N. and Meyer, C. D. (2006). *Google's PageRank and beyond: The Science of Search Engine Rankings* (Princeton University Press, Princeton).

Lee, S., Rocha, L. E. C., Liljeros, F. and Holme, P. (2012). Exploiting temporal network structures of human interaction to effectively immunize populations, *PLOS ONE* **7**, article No. e36439.

Lusseau, D., Schneider, K., Boisseau, O. J., Haase, P., Slooten, E. and Dawson, S. M. (2003). The bottlenose dolphin community of Doubtful Sound features a large proportion of long-lasting associations. *Behavioral Ecology and Sociobiology* **54**, pp. 396–405.

Liben-Nowell, D. and Kleinberg, J. (2007). The link-prediction problem for social networks, *Journal of the American Society of Information Science and Technology* **58**, pp. 1019–1031.

Liberzon, D. (2003). *Switching in Systems and Control* (Springer Science + Business Media, New York).

Liu, S., Perra, N., Karsai, M. and Vespignani, A. (2014). Controlling contagion processes in activity driven networks, *Physical Review Letters* **112**, article No. 118702.

Liu, S. -Y., Baronchelli, A. and Perra, N. (2013). Contagion dynamics in time-varying metapopulation networks, *Physical Review E* **87**, article No. 032805.

Lloyd, A. L. (2001). Destabilization of epidemic models with the inclusion of realistic distributions of infectious periods, *Proceedings of the Royal Society of London B* **268**, pp. 985–993.

Lü, L. and Zhou, T. (2011). Link prediction in complex networks: A survey, *Physica A* **390**, pp. 1150–1170.

Lusher, D., Koskinen, J. and Robins, G. (eds.) (2013). *Exponential Random Graph Models for Social Networks — Theory, Methods, and Applications* (Cambridge University Press, Cambridge).

Mahmoud, H. M., Smythe, R. T. and Szymański, J. (1993). On the structure of random plane-oriented recursive trees and their branches, *Random Structures and Algorithms* **4**, pp. 151–176.

Malmgren, R. D., Stouffer, D. B., Campanharo, A. S. L. O. and Amaral, L. A. N. (2009). On universality in human correspondence activity, *Science* **325**, pp. 1696–1700.

Malmgren, R. D., Stouffer, D. B., Motter, A. E. and Amaral, L. A. N. (2008). A Poissonian explanation for heavy tails in e-mail communication, *Proceedings of the National Academy of Sciences of the United States of America* **105**, pp. 18153–18158.

Mantzaris, A. V. and Higham, D. J. (2013). Dynamic communicability predicts infectiousness, In: Holme, P. and Saramäki, J. (eds.), *Temporal Networks* (Springer, Berlin), pp. 283–294.

Masuda, N. (2010). Effects of diffusion rates on epidemic spreads in metapopulation networks, *New Journal of Physics* **12**, article No. 093009.

Masuda, N. and Holme, P. (2013). Predicting and controlling infectious disease epidemics using temporal networks, *F1000Prime Reports* **5**, article No. 6.

Masuda, N., Kim, J. S. and Kahng, B. (2009). Priority queues with bursty arrivals of incoming tasks, *Physical Review E* **79**, article No. 036106.

Masuda, N., Klemm, K. and Eguíluz, V. M. (2013a). Temporal networks: Slowing down diffusion by long lasting interactions, *Physical Review Letters* **111**, article No. 188701.

Masuda, N. and Rocha, L. E. C. (2016). A Gillespie algorithm for non-Markovian stochastic processes: Laplace transform approach, Preprint: arXiv:1601.01490v1.

Masuda, N., Takaguchi, T., Sato, N. and Yano, K. (2013b). Self-exciting point process modeling of conversation event sequences, In: Holme P. and Saramäki, J. (eds.), *Temporal Networks* (Springer-Verlag, Berlin), pp. 245–264.

Mattern, F. (1989). Virtual time and global states of distributed systems, *Proceedings of Workshop on Parallel and Distributed Algorithms*, pp. 215–226.

mfinder   (2016).   https://www.weizmann.ac.il/mcb/UriAlon/download/network-motif-software

Milo, R., Itzkovitz, S., Kashtan, N., Levitt, R., Shen-Orr, S., Ayzenshtat, I., Sheffer, M. and Alon, U. (2004). Superfamilies of evolved and designed networks, *Science* **303**, pp. 1538–1542.

Milo, R., Shen-Orr, S., Itzkovitz, S., Kashtan, N., Chklovskii, D. and Alon, U. (2002). Network motifs: Simple building blocks of complex networks, *Science* **298**, pp. 824–827.

Min, B., Goh, K. -I. and Kim, I. -M. (2009). Waiting time dynamics of priority-queue networks, *Physical Review E* **79**, article No. 056110.

Montroll, E. W. and Weiss, G. H. (1965). Random walks on lattices. II, *Journal of Mathematical Physics* **6**, pp. 167–181.

Moreau, L. (2005). Stability of multiagent systems with time-dependent communication links, *IEEE Transactions on Automatic Control* **50**, pp. 169–182.

Motegi, S. and Masuda, N. (2012). A network-based dynamical ranking system for competitive sports, *Scientific Reports* **2**, article No. 904.

Mucha, P. J., Richardson, T., Macon, K., Porter, M. A. and Onnela, J. -P. (2010). Community structure in time-dependent, multiscale, and multiplex networks, *Science* **328**, pp. 876–878.

Newman, M. E. J. (2002). Spread of epidemic disease on networks, *Physical Review E* **66**, article No. 016128.

Newman, M. E. J. (2004). Fast algorithm for detecting community structure in networks, *Physical Review E* **69**, article No. 066133.

Newman, M. E. J. (2005). Power laws, Pareto distributions and Zipf's law, *Contemporary Physics* **46**, pp. 323–351.

Newman, M. E. J. (2010). *Networks — An Introduction* (Oxford University Press, Oxford).

Newman, M. E. J., Strogatz, S. H. and Watts, D. J. (2001). Random graphs with arbitrary degree distributions and their applications, *Physical Review E* **64**, article No. 026118.

Nicosia, V., Tang, J., Musolesi, M., Russo, G., Mascolo, C. and Latora, V. (2012). Components in time-varying graphs, *Chaos* **22**, article No. 023101.

Noh, J. D., Shim, G. M. and Lee, H. (2005). Complete condensation in a zero range process on scale-free networks, *Physical Review Letters* **94**, article No. 198701.

Ogata, Y. (1988). Statistical models for earthquake occurrences and residual analysis for point processes, *Journal of the American Statistical Association* **83**, pp. 9–27.

Ogata, Y. (1999). Seismicity analysis through point-process modeling: A review, *Pure and Applied Geophysics* **155**, pp. 471–507.

Olfati-Saber, R., Fax, J. A. and Murray, R. M. (2007). Consensus and cooperation in networked multi-agent systems, *Proceedings of the IEEE* **95**, pp. 215–233.

Olfati-Saber, R. and Murray, R. M. (2004). Consensus problems in networks of agents with switching topology and time-delays, *IEEE Transactions on Automatic Control* **49**, pp. 1520–1533.

Ozaki, T. (1979). Maximum likelihood estimation of Hawkes' self-exciting point processes, *Annals of the Institute of Statistical Mathematics* **31**, pp. 145–155.

Palla, G., Barabási, A. -L. and Vicsek, T. (2007). Quantifying social group evolution, *Nature* **446**, pp. 664–667.

Palla, G., Derényi, I., Farkas, I. and Vicsek, T. (2005). Uncovering the overlapping community structure of complex networks in nature and society, *Nature* **435**, pp. 814–818.

Pan, R. K. and Saramäki, J. (2011). Path lengths, correlations, and centrality in temporal networks, *Physical Review E* **84**, article No. 016105.

Park, J. and Newman, M. E. J. (2005). A network-based ranking system for US college football, *Journal of Statistical Mechanics*, article No. P10014.

Pastor-Satorras, R., Castellano, C., Van Mieghem, P. and Vespignani, A. (2015). Epidemic processes in complex networks, *Reviews of Modern Physics* **87**, pp. 925–979.

Peel, L. and Clauset, A. (2015). Detecting change points in the large-scale structure of evolving networks, *Proceedings of the Twenty-ninth AAAI Conference on Artificial Intelligence*, pp. 2914–2920.

Peixoto, T. P. and Rosvall, M. (2015). Modeling sequences and temporal networks with dynamic community structures, Preprint arXiv:1509.04740v1.

Peng, C. -K., Buldyrev, S. V., Havlin, S., Simons, M., Stanley, H. E. and Goldberger, A. L. (1994). Mosaic organization of DNA nucleotides, *Physical Review E* **49**, pp. 1685–1689.

Pernice, V., Staude, B., Cardanobile, S. and Rotter, S. (2011). How structure determines correlations in neuronal networks, *PLOS Computational Biology* **7**, article No. e1002059.

Perra, N., Baronchelli, A., Mocanu, D., Gonçalves, B., Pastor-Satorras, R. and Vespignani, A. (2012a). Random walks and search in time-varying networks, *Physical Review Letters* **109**, article No. 238701.

Perra, N., Gonçalves, B., Pastor-Satorras, R. and Vespignani, A. (2012b). Activity driven modeling of time varying networks, *Scientific Reports* **2**, article No. 469.

Pfitzner, R., Scholtes, I., Garas, A., Tessone, C. J. and Schweitzer, F. (2013). Betweenness preference: Quantifying correlations in the topological dynamics of temporal networks, *Physical Review Letters* **110**, article No. 198701.

Pincombe, B. (2005). Anomaly detection in time series of graphs using ARMA processes, *ASOR Bulletin* **24**, December Issue, pp. 1–10.

Porfiri, M., Stilwell, D. J., Bollt, E. M. and Skufca, J. D. (2006). Random talk: Random walk and synchronizability in a moving neighborhood network, *Physica D* **224**, pp. 102–113.

Porter, M. A. and Gleeson, J. P. (2014). Dynamical systems on networks: A tutorial. Preprint: arXiv:1403.7663v2.

Ren, W. and Beard, R. W. (2005). Consensus seeking in multiagent systems under dynamically changing interaction topologies, *IEEE Transactions on Automatic Control* **50**, pp. 655–661.

Ribeiro, B., Perra, N. and Baronchelli, A. (2013). Quantifying the effect of temporal resolution on time-varying networks, *Scientific Reports* **3**, article No. 3006.

Rizzo, A., Frasca, M. and Porfiri, M. (2014). Effect of individual behavior on epidemic spreading in activity-driven networks, *Physical Review E* **90**, article No. 042801.

Robins, G. and Pattison, P. (2001). Random graph models for temporal processes in social networks, *Journal of Mathematical Sociology* **25**, pp. 5–41.

Robins, G., Pattison, P., Kalish, Y. and Lusher, D. (2007). An introduction to exponential random graph (p^*) models for social networks, *Social Networks* **29**, pp. 173–191.

Rocha, L. E. C., Liljeros, F. and Holme, P. (2011). Simulated epidemics in an empirical spatiotemporal network of 50,185 sexual contacts, *PLOS Computational Biology* **7**, article No. e1001109.

Rocha, L. E. C. and Masuda, N. (2014). Random walk centrality for temporal networks, *New Journal of Physics* **16**, article No. 063023.

Rosvall, M. and Bergstrom, C. T. (2008). Maps of random walks on complex networks reveal community structure, *Proceedings of the National Academy of Sciences of the United States of America* **105**, pp. 1118–1123.

Rosvall, M. and Bergstrom, C. T. (2010). Mapping change in large networks, *PLOS ONE* **5**, article No. e8694.

Rosvall, M., Esquivel, A. V., Lancichinetti, A., West, J. D. and Lambiotte, R. (2014). Memory in network flows and its effects on spreading dynamics and community detection, *Nature Communications* **5**, article No. 4630.

Rybski, D., Buldyrev, S. V., Havlin, S., Liljeros, F. and Makse, H. A. (2012). Communication activity in a social network: Relation between long-term correlations and inter-event clustering, *Scientific Reports* **2**, article No. 560.

Saramäki, J. and Holme, P. (2015). Exploring temporal networks with greedy walks, *European Physical Journal B* **88**, article No. 334.

Scholtes, I., Wider, N., Pfitzner, R., Garas, A., Tessone, C. J. and Schweitzer, F. (2014). Causality-driven slow-down and speed-up of diffusion in non-Markovian temporal networks, *Nature Communications* **5**, article No. 5024.

Schreiber, T. (2000). Measuring information transfer, *Physical Review Letters* **85**, pp. 461–464.

Shinomoto, S., Shima, K. and Tanji, J. (2003). Differences in spiking patterns among cortical neurons, *Neural Computation* **15**, pp. 2823–2842.

Skiena, S. S. (2008). *The Algorithm Design Manual, Second Edition* (Springer, London).

Smith, W. L. (1961). A note on the renewal function when the mean renewal lifetime is infinite, *Journal of the Royal Statistical Society Series B* **23**, pp. 230–237.

Snijders, T. A. B. (2001). The statistical evaluation of social network dynamics, *Sociological Methodology* **31**, pp. 361–395.

So, P., Cotton, B. C. and Barreto, E. (2008). Synchronization in interacting populations of heterogeneous oscillators with time-varying coupling, *Chaos* **18**, article No. 037114.

Softky, W. R. and Koch, C. (1993). The highly irregular firing of cortical cells is inconsistent with temporal integration of random EPSPs, *Journal of Neuroscience* **13**, pp. 334–350.

Song, C., Qu, Z., Blumm, N. and Barabási, A. -L. (2010). Limits of predictability in human mobility, *Science* **327**, pp. 1018–1021.

Sornette, D., Deschâtres, F., Gilbert, T. and Ageon, Y. (2004). Endogenous versus exogenous shocks in complex networks: An empirical test using book sale rankings, *Physical Review Letters* **93**, article No. 228701.

Speidel, L., Klemm, K., Eguíluz, V. M. and Masuda, N. (2016). Temporal interactions facilitate endemicity in the susceptible-infected-susceptible epidemic model, Preprint: arXiv:1602.00859v1.

Speidel, L., Lambiotte, R., Aihara, K. and Masuda, N. (2015). Steady state and mean recurrence time for random walks on stochastic temporal networks, *Physical Review E* **91**, article No. 012806.

Starnini, M., Baronchelli, A., Barrat, A. and Pastor-Satorras, R. (2012). Random walks on temporal networks, *Physical Review E* **85**, article No. 056115.

Starnini, M. and Pastor-Satorras, R. (2013). Topological properties of a time-integrated activity-driven network, *Physical Review E* **87**, article No. 062807.

Starnini, M. and Pastor-Satorras, R. (2014). Temporal percolation in activity-driven networks, *Physical Review E* **89**, article No. 032807.

Stehlé, J., Voirin, N., Barrat, A., Cattuto, C., Isella, L., Pinton, J. -F., Quaggiotto, M., Van den Broeck, W., Régis, C., Lina, B. and Vanhems, P. (2011). High-resolution measurements of face-to-face contact patterns in a primary school, *PLOS ONE* **6**, article No. e23176.

Stilwell, D. J., Bollt, E. M. and Roberson, D. G. (2006). Sufficient conditions for fast switching synchronization in time-varying network topologies, *SIAM Journal of Applied Dynamical Systems* **5**, pp. 140–156.

Sugihara, G., May, R., Ye, H., Hsieh, C. -H., Deyle, E., Fogarty, M. and Munch, S. (2012). Detecting causality in complex ecosystems, *Science* **338**, pp. 496–500.

Szymański, J. (1987). On a nonuniform random recursive tree, *Annals of Discrete Mathematics* **33**, pp. 297–306.

Tabourier, L., Libert, A. -S. and Lambiotte, R. (2016). Predicting links in ego-networks using temporal information, *EPJ Data Science* **5**, article No. 1.

Takaguchi, T. and Masuda, N. (2011). Voter model with non-Poissonian inter-event intervals, *Physical Review E* **84**, article No. 036115.

Takaguchi, T., Masuda, N. and Holme, P. (2013). Bursty communication patterns facilitate spreading in a threshold-based epidemic dynamics, *PLOS ONE* **8**, article No. e68629.

Takaguchi, T., Nakamura, M., Sato, N., Yano, K. and Masuda, N. (2011). Predictability of conversation partners, *Physical Review X* **1**, article No. 011008.

Takaguchi, T., Sato, N., Yano, K. and Masuda, N. (2012). Importance of individual events in temporal networks, *New Journal of Physics* **14**, article No. 093003.

Tang, J., Musolesi, M., Mascolo, C. and Latora, V. (2009). Temporal distance metrics for social network analysis, *Proceedings of the 2nd ACM Workshop on Online Social Networks (WOSN)*, pp. 31–36.

Tang, J., Musolesi, M., Mascolo, C., Latora, V. and Nicosia, V. (2010a). Analysing information flows and key mediators through temporal centrality metrics, *Proceedings of the 3rd Workshop on Social Network Systems (SNS)*, article No. 3.

Tang, J., Scellato, S., Musolesi, M., Mascolo, C. and Latora, V. (2010b). Small-world behavior in time-varying graphs, *Physical Review E* **81**, article No. 055101(R).

Taqqu, M. S., Teverovsky, V. and Willinger, W. (1995). Estimators for long-range dependence: An empirical study, *Fractals* **3**, pp. 785–798.

Taylor, D., Myers, S. A., Clauset, A., Porter, M. A. and Mucha, P. J. (2015). Eigenvector-based centrality measures for temporal networks, Preprint: arXiv:1507.01266v1.

Teugels, J. L. (1968). Renewal theorems when the first or the second moment is infinite, *Annals of Mathematical Statistics* **39**, pp. 1210–1219.

Tokdar, S., Xi, P., Kelly, R. C. and Kass, R. E. (2010). Detection of bursts in extracellular spike trains using hidden semi-Markov point process models, *Journal of Computational Neuroscience* **29**, pp. 203–212.

Tremblay, N. and Borgnat, P. (2014). Graph wavelets for multiscale community mining, *IEEE Transactions on Signal Processing* **62**, pp. 5227–5239.

Vázquez, A., Oliveira, J. G., Dezsö, Z., Goh, K.-I., Kondor, I. and Barabási, A.-L. (2006). Modeling bursts and heavy tails in human dynamics, *Physical Review E* **73**, article No. 036127.

Vazquez, A., Rácz, B., Lukács, A. and Barabási, A. -L. (2007). Impact of non-Poissonian activity patterns on spreading processes, *Physical Review Letters* **98**, article No. 158702.

Vere-Jones, D. (1970). Stochastic models for earthquake occurrence, *Journal of the Royal Statistical Society Series B* **32**, pp. 1–62.

Volz, E. (2008). SIR dynamics in random networks with heterogeneous connectivity, *Journal of Mathematical Biology* **56**, pp. 293–310.

Volz, E. and Meyers, L. A. (2007). Susceptible-infected-recovered epidemics in dynamic contact networks, *Proceedings of the Royal Society B* **274**, pp. 2925–2933.

Wasserman, S. and Pattison, P. (1996). Logit models and logistic regressions for social networks: I. An introduction to Markov graphs and p^*, *Psychometrika* **61**, pp. 401–425.

Watts, D. J. and Strogatz, S. H. (1998). Collective dynamics of 'small-world' networks, *Nature* **393**, pp. 440–442.

Xu, K. S. and Hero III, A. O. (2014). Dynamic stochastic blockmodels for time-evolving social networks, *IEEE Journal of Selected Topics in Signal Processing* **8**, pp. 552–562.

Xuan, B. B., Ferreira, A. and Jarry, A. (2003). Computing shortest, fastest, and foremost journeys in dynamic networks, *International Journal of Foundations of Computer Science* **14**, pp. 267–285.

Yang, T., Chi, Y., Zhu, S., Gong, Y. and Jin, R. (2011). Detecting communities and their evolutions in dynamic social networks — A Bayesian approach, *Machine Learning* **82**, pp. 157–189.

Yook, S. H., Jeong, H., Barabási, A. -L. and Tu, Y. (2001). Weighted evolving networks, *Physical Review Letters* **86**, pp. 5835–5838.

Index

CPSIA information can be obtained
at www.ICGtesting.com
Printed in the USA
LVHW010054270321
682643LV00001B/50